CLEAN COAL/DIRTY AIR

CLEAN COAL/DIRTY AIR

or How the Clean Air Act Became
a Multibillion-Dollar Bail-Out for
High-Sulfur Coal Producers and
What Should Be Done about It

BRUCE A. ACKERMAN
and
WILLIAM T. HASSLER

New Haven and London
Yale University Press

A preliminary version of this work was published by the *Yale Law Journal*
in its July 1980 issue under the title "Beyond the New Deal: Coal and
the Clean Air Act." Copyright to this version is owned by
Susan Rose-Ackerman, Trustee.

Designed by Nancy Ovedovitz and set in VIP Baskerville type.
Printed in the United States of America by Vail-Ballou Press, Binghamton, N.Y.

Library of Congress Cataloging in Publication Data

Ackerman, Bruce A.
   Clean coal/dirty air.

   "Preliminary version . . . published by the Yale law journal in its July 1980
issue under the title 'Beyond the New Deal: coal and the Clean air act.' "
   Includes index.
   1. Air—Pollution—Law and legislation—United States.   2. Coal—Law and
legislation—United States.   I. Hassler, William T.   II. Title.
KF3812.A93   344.73′046342   80-54219
                                       AACR2
ISBN 0-300-02628-5
ISBN 0-300-02643-9 (pbk.)

10 9 8 7 6 5 4 3 2 1

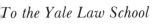
To the Yale Law School

CONTENTS

ACKNOWLEDGMENTS

Work on this essay began in May 1979, when Bill Hassler accompanied Bruce Ackerman for the latter's two-week stay as visiting professor with the General Counsel's Office of the Environmental Protection Agency in Washington, D.C. This was a time of intensive interviewing with individuals both in and out of government. Summer and fall were spent studying the docket and supplementing the initial set of interviews. We are especially grateful to the more than fifty people who, under pledge of confidentiality, spoke with us. Pursuant to our commitments to them, we have not attributed particular findings to particular interviewees. Nonetheless, we could not have written the book without them.

During the past year, our work has been supported by grants from the Kaiser and Commonwealth Foundations to Yale's Center for Health Studies, a part of the university's Institution for Social and Policy Studies (ISPS). We especially appreciate the moral and intellectual support provided by the participants in the ISPS faculty seminar on health policy led by Ted Marmor. No less important was the aid forthcoming from the Yale Law School. A triennial leave permitted Bruce Ackerman to work uninterruptedly on this essay during the spring of 1980. And Dean Harry Wellington supported the project with a tact and intelligence too easily taken for granted.

We are also greatly indebted to the editorial board of the *Yale Law Journal*, who published an earlier form of this study under the title "Beyond the New Deal: Coal and the Clean Air Act" in their July 1980 issue. In the process of preparing the article for publication, the *Journal* editors were incredibly conscientious in their efforts to refine our arguments and check our sources—while nobody was punching time clocks, it would not surprise us if the *Journal* invested more than a thousand person-hours in the enterprise. We are especially grateful to the articles editors responsible for smoothing our path to print: Carol Lee, Dwight Smith, and Ken Taymor. This book presents new arguments and data that did not appear in the article, but it builds upon the earlier study to which the *Journal* staff contributed so much. We also relied greatly on Linda Larson and Gabrielle Simonson, who typed what must have seemed an endless stream of revisions with extraordinary ability and good will.

And finally, there are the people who helped us along the way. Bill Hassler is grateful for the opportunity to publicly acknowledge his debt to his parents, William and Helen Hassler, and to his aunt, Dot Sims. Bruce Ackerman simply notes the extent to which this work flows from fifteen years of conversation with Susan Rose-Ackerman.

1

BEYOND THE NEW DEAL

The passions of Earth Day have marked our law in deep and abiding ways. Statutes passed in the early 1970s did more than commit hundreds of billions of dollars to the cause of environmental protection in the decades ahead.[1] They also represent part of a complex effort by which the present generation is revising the system of administrative law inherited from the New Deal. The rise of environmental consciousness in the late 1960s coincided with the decline of an older dream—the image of an independent and expert administrative agency creatively regulating a complex social problem in the public interest. When Congress reacted to Earth Day, it tried to do more than clean the water and purify the air; it also sought a new shape for the administrative process—one that would avoid the use of expertise as an excuse for inaction and would protect agencies from capture by special interests. It is a decade now since Congress began to articulate this new vision of administrative law—long enough for us to begin to test its aspirations against concrete results. In this spirit, we propose to sift a decade's experience generated by one of the countless experiments in administrative lawmaking written into the Clean Air Act of 1970.[2] We seek to understand how decision makers perceived, defined, and solved problems within the evolving framework of environmental regulation—so that we may begin to distinguish experiments in administrative design that have succeeded from those that have failed.

Beyond this, our study also focuses upon a crucial substantive policy issue: the future of the coal-burning power plant. At present, these plants contribute 48 percent of all electric power produced in the United States.[3] This share will grow over the next half century.[4] With oil scarce, nuclear risky, solar embryonic, and hydro limited, the nation's rich and cheap coal reserves call for exploitation.[5] At the same time, coal burning generates environmental burdens. Coal-fired power plants are major sources of several pollutants; they are currently the single most important source of sulfur oxides.[6] As a result, the control of new coal burners has gradually emerged as one of the most pressing questions confronting the Environmental Protection Agency (EPA), leading it, in June 1979, to revise the "new source performance standards" (NSPS) it had previously imposed on sulfur emissions from new coal-burning power plants.[7] Environmentalists and the utility industry have challenged the new NSPS in litigation currently before the United States Court of Appeals for the District of Columbia Circuit.[8]

A careful analysis of the EPA's attempt to control coal burning will reveal the way our institutions are resolving a critical environmental trade-off generated by the energy crisis. We have concluded that Congress's well-intentioned effort to improve the administrative process has driven EPA to an extraordinary decision that will cost the public tens of billions of dollars to achieve environmental goals that could be reached more cheaply, more quickly, and more surely by other means. Indeed, the agency action is so inept that some of the nation's most populous areas will end up with a *worse* environment than would have resulted if the new policy had never been put into effect. Yet the people who shaped the 1979 decision are remarkable for their high intelligence and conscientiousness. Their failure to make sensible policy is a symptom of organizational, not personal, breakdown—a failure to give decision makers bureaucratic incentives to ask the hard questions raised by any serious effort to control the environment. Thus, to understand the decision, we must do more than outline the substance of the environmental problem and the administrative response. We must address broader questions raised by the framework within which Congress, agencies, courts, and special interest groups interacted to form and implement policy.

Our story begins with the way Congress set about to control the environment in 1970. Rather than consign the regulation of new power plants to an independent expert agency of the kind idealized by New Deal theory, Congress tried to play a more aggressive role in policymaking. There are many different ways, however, that Congress could have tried to direct agency policy; indeed, it will be a principal purpose of this study to isolate the *agency-forcing* statute as a legal form worthy of careful analysis. Yet in its eagerness to move beyond New Deal ideals, the Congress of 1970 hit upon a form of policy intervention that replaced familiar New Deal maladies with new, but hardly less serious, diseases. EPA responded to Congress's primitive effort at agency-forcing by creating a regulatory universe only tangentially related to environmental realities. It treated the power plant problem as if it were an engineering exercise insulated from critical ecological and economic issues.

After setting the bureaucratic stage, we trace the way the coal burner rose to the surface of congressional deliberations in 1976 and 1977, when the legislature was obliged to confront the consequences of its earlier exercises in aggressive policymaking. We then compare this post–New Deal process of congressional amendment with a viewpoint commended by New Deal ideals—one that focuses on the relationship between economic costs and ecological benefits promised by alternative regulatory policies. Having contrasted the competing decision-making perspectives, we can begin to appreciate the perplexities the EPA confronted in 1977, when it attempted to place its expertise at the service of a new congressional exercise in agency-forcing enacted as part of the Clean Air Act amendments of 1977. At this point, the conflict between administrative philosophies took institutional form, with different parts of the executive bureaucracy reacting differently to the congressional effort to move beyond the New Deal. It is only in the light of this bureaucratic struggle that the EPA's 1979 decision on new standards for coal burners becomes understandable, if not rational.

Our study concludes by exploring the larger implication of the EPA's decision for the future of environmental law. We try to identify an appropriate role for courts called upon to resolve the perplexities generated by agency-forcing. We then shift focus from judicial review to legislative reform. Congress was not wrong in hoping to move beyond New Deal ideals in 1970; it is imperative, however, that its early efforts in agency-forcing be replaced by statutory structures that promise a more fruitful dialogue between politicians and technocrats in the decades ahead.

THE NEW DEAL IDEAL

Imagine that Earth Day had fallen a generation earlier—when Franklin Roosevelt, rather than Richard Nixon, was in the White House. In response to a sense of crisis, the president asserts that the environment deserves a New Deal and turns to his Brain Trust for a legislative remedy. How would the New Deal have sought to transform the environmental impulse into a legal reality?

We shall isolate three distinct, if interrelated, elements of the New Deal answer.[9] The first is the *affirmation of expertise*. Understanding the environment, after all, is an exceedingly technical business requiring the coordination of a bewildering variety of specialties ranging from ecology to engineering to economics. The central New Deal mission is to create a decision-making structure capable of deploying the varieties of relevant expert knowledge. Without a sober understanding of the scientific and social facts, there can be no hope of defining an intelligent solution to the chronic problems of a complex and interdependent society. In affirming the need for expertise, however, good New Dealers should not be blind to the failures of particular experts. Most important, particular people can lose touch with the current of informed opinion: science marches on, new facts are found, new theories are proposed. While an administrative agency may not be at the frontier of research, the very least it should do is to keep up with the consensus of expert opinion. This means that the ideal agency may, from time to time, be required to change course dramatically to take into account new insights generated by the best available knowledge. Such flexibility is absolutely essential if the New Deal agency is to free itself from ideas that no longer illuminate the changing reality it seeks to regulate.

Two institutional corollaries follow from the affirmation of expertise. The first is *agency insulation from central political control*. At the very best, Congress will freeze the existing expert consensus into law, thereby making it harder to cope with the problem of obsolescence. Moreover, there is little reason to expect the very best. In the words of James Landis, the most thoughtful New Deal theorist, "those with experience in legislative matters . . . recognize that it is easier to plot a way through a labyrinth of detail when it is done in the comparative quiet of [an agency] conference room than when it is attempted amid the turmoil of a legislative chamber or committee room."[10] Rather than tie the agency's hands with a host of particular rules and detailed instructions, Congress should content itself with the most general kinds of policy guidance. By restricting itself to the role of Polonius, Congress gives evidence of a self-conscious awareness of its institutional incompetence. Instead of imposing a hard and fast solution to a complex and changing problem, the legislature should invite the agency to organize the expert knowledge required for intelligent regulation.

To implement this goal, Congress should try to insulate the agency from other sources of power that might readily overwhelm its deepening understanding of its policy problem. By making the agency independent from the executive[11] and by endowing it with multiple commissioners,[12] the New Deal makes it difficult for a momentary national impulse to affect agency policy. An even more extreme form of insulation is provided by "cooperative federalism." Here the states, operating under loose federal supervision, are asked to design a program responsive to the peculiarities of local conditions.[13]

Not that insulation is ever conceived as complete. Sustained shifts in national values will be expressed over time in agency appointments and appropriations. Agency policymaking, however, is not to be subordinated to every blip in national opinion; first and foremost, the agency is to be responsive to the evolving complexity of the substantive problem it confronts.

But the New Deal agency is not only to be insulated from national politics; it is also to be *insulated from judicial oversight*. The overriding aim of administrative law is to discourage courts from displacing expert policy judgments by their own legalistic readings of the statute.[14] Rather than undermining New Deal aspirations, judicial review should focus on questions that promise to support the use of expertise—most notably, did the agency give serious consideration to all the relevant data and arguments?[15] Beyond this, the courts are to restrain the temptation to second-guess agency decisions, striking down only those rare cases of arbitrary or capricious action which patently belie the myth of expertise.[16]

If Earth Day had fallen in the late 1930s, then, one might imagine the Brain Trust advancing a familiar legislative remedy to a new social dis-ease. Congress should create an independent agency consisting of five commissioners (no more than three from the same party) to take on the enormous regulatory task. After sagely instructing the commissioners to preserve the integrity of nature, the health of the citizenry, and the prosperity of the economy, the statute should leave the commission on its own as it seeks to define a sound environmental policy. Rather than giving concrete policy guidance, the statute should diffuse political direction still further by delegating vast areas of policymaking to the states.

Legalism should be kept to the periphery. Doubtless, there must be some guarantee that interested parties can obtain a hearing before the commissioners impose clean-up burdens upon them. Doubtless industry will try to tie the commission's hands by making these hearings as long and complicated (and as courtlike) as possible. Yet the courts should support the expert agency's effort to resist these tendencies—giving affected parties their constitutional due without depriving the agency of policy-making flexibility.[17]

A NEW GENERATION CONFRONTS THE NEW DEAL

But Earth Day did not fall during the Great Depression. By the late 1960s, a generation's experience had eroded New Deal confidence in expert policymaking. Two themes could be detected amid the emerging chorus of criticism.[18] The first, more conservative, line of attack did not directly assault the New Deal affirmation of expertise. Instead, it asserted that the agencies had somehow failed to make use of their broad rule-making powers to engage in creative policymaking in the public interest. They had relapsed into the old lawyer-ridden ways of case-by-case adjudication, laboring mightily through procedural labyrinths without successfully defining basic directions for future regulation. Rather than becoming a home for dedicated experts, the independent commission seemed to be a revolving door for lawyers hoping to gain inside experience that could later be cashed out in lucrative private practice. Instead of encouraging an impartial search for the public interest, the collegial structure of the independent agency mired would-be policymakers in collective indecision.[19] These criticisms of agency performance merged, sometimes imperceptibly, into a more radical critique of the New Deal ideal itself. On this line of attack, expertise was seen as a myth concealing the inevitability of hard value choices, and agency insulation from political or judicial oversight as a screen concealing the capture of the agency by special interests.[20]

These criticisms generated a predictable set of proposed remedies. On the one hand, there was an increasingly impatient demand that the agencies finally redeem their New Deal promise—to establish clear standards through creative rule making.[21] On the other hand, there was an increasing temptation to tinker with the institutional corollaries associated with the New Deal ideal. If existing agencies did not redeem New Deal ideals, perhaps some creative legislative or judicial responses would make a difference.[22] Responding to the prevailing sense of unease, both courts and legislatures began experimenting with a wide variety of different—sometimes contradictory—solutions that promised to fill the legitimacy gap that had opened before them.[23]

Environmental law proved to be especially fertile ground for these proliferating experiments.[24] When legal activists tried to give their environmental hopes statutory expression in the early 1970s, their concrete experiences gave added weight to the growing suspicion of New Deal models among the American establishment.[25] Before 1970, environmental protection was principally a matter for the states rather than the federal government. And when environmentalists surveyed the state scene, the agencies they observed seemed to parody New Deal hopes. The typical state agency was so understaffed that it could not even pretend to understand the environment it was trying to regulate. Although state agencies frequently took the form of "independent" commissions, their memberships were often dominated by the very interests that had the most to gain from pollution.[26]

In response to this dismal reality, the Clean Air Amendments of 1970[27] not only massively increased the federal presence but took steps to guard against the repetition of yet another New Deal failure.[28] Instead of permitting a group of "independent" commissioners to run off in different directions, the act placed primary responsibility on a single administrator squarely situated within the executive branch.[29] Moreover, to ensure that the administrator implemented the act fully, Congress permitted "any citizen" to sue to compel the agency to carry out the act's mandatory sections.[30]

Just as the act refused to insulate the EPA in New Deal fashion, so too it challenged the New Deal affirmation of expertise in two very different ways. First, the act not only required the administrator to set quantitative clean air targets that "protect the public health,"[31] but it also insisted that the nation actually fulfill these clean air targets by 1977 at the latest.[32] In taking these steps, Congress forced the agency to specify its goals far more clearly than required by the New Deal model. No longer could expertise be used as an excuse for avoiding the inevitably controversial task of defining ultimate environmental objectives; instead the agency had to define its goals in a highly visible way and recognize that Congress would call it to account by a specific date if it found the agency's performance unsatisfactory.

At the same time it energetically pursued this *ends-forcing* strategy, Congress experimented with another form of agency-forcing in a more tentative way. Once having set air quality targets, the next step was to define the best means of achieving the pollution cutbacks that would be required to clean the air by 1977. At this stage, Congress was more reluctant to make a total break with New Deal models. Indeed, as far as plants that existed in 1970 were concerned, the act remitted the problem of defining individual clean-up obligations to a classical New Deal process in which the states were to play a leading role. Apprised of the federal clean air targets, the states were obliged to proffer a State Implementation Plan (SIP) that would detail the way they would achieve the ambient objectives by the congressional target date. In implementing the SIP, local officials—with federal assistance—tried to build mathematical models that estimated the impact of important dischargers on local air quality. Effluent limits were then imposed by local administrators through a low visibility, and highly discretionary, process.[33] The binding federal constraint was that, taken together, polluters within each airshed had to reduce emissions sufficiently to bring local conditions into compliance with federal clean air targets.[34] Since different airsheds had different air quality and contained different polluters with different clean-up costs, different SIPs required old plants to reduce their discharges in widely varying amounts. To take an example central to the discussion, present SIPs impose vastly different sulfur oxide limitations on old coal burners. While some SIPs limit plants to .8 pounds of sulfur dioxide for every million BTUs (MBTU) of energy they produce, many others, especially in the Midwest, allow plants to emit 4 or 5 pounds of sulfur dioxide to produce the same amount of energy.[35]

Nor is this disparity a bad thing—at least when viewed through New Deal eyes. Since the pollution problem is of different severity in different parts of the country, why should different polluters not be asked to cut back accordingly? If, due to local conditions, it is relatively expensive for coal burners to cut back their emissions compared to other dischargers, why should the SIPs not take this into account? In principle, the decentralized SIP process could be structured in a way that permitted knowledgeable experts to work effectively for the public good—designing cut-back requirements to meet the EPA air quality goals in a fair and efficient manner.[36]

In contrast, the act's treatment of *new* plants represented a sharper break with New Deal ideals. Rather than integrating new plants into the SIP process, the control responsibilities of new coal burners would be determined in an entirely different way. No longer would there be a self-conscious effort to tailor cut-back responsibilities to local conditions; instead, the act's approach to New Source Performance Standards required all plants of the same type, regardless of their location, to meet the same emission ceiling.[37] Equally important, the administrator was deprived of New Deal policymaking space when setting the NSPS. Rather than instructing EPA to set a limitation that "protected the environment while assuring a vigorous economy," the statute tried to define the clean-up responsibilities of new plants in a decisive way. Section 111 (as originally written) required the administrator to set an effluent standard that could be satisfied by the "best system of emission reduction which (taking into account the costs of achieving such reduction) the Administrator determines has been adequately demonstrated."[38] While this formula was fairly elastic, its general thrust was plain. So far as new plants were concerned, Congress not only would force the EPA to specify its *ends* with clarity, but also presumed to specify the best *means* of achieving clean air objectives: each and every new plant would be required to install the "best system of emission reduction."

Up to a point, this insistence on better performance from new plants made good sense. Old plants, after all, had often been designed with little or no thought to pollution control, and the new limitations would frequently require expensive retrofitting. In contrast, new plants could be designed from the start to take pollution reduction into account. By lifting NSPS out of the general effort to adapt cut-back requirements to local environmental conditions, though, the Clean Air Act made it easier for policymakers to push NSPS beyond the point of ecological justification.[39]

From the vantage point of the environment,* it makes no difference whether a pound of sulfur oxide is emitted by a power plant built before or after some magic date. Nor does it matter whether the sulfur is removed by primitive hosing or by the most advanced gismo ever conceived in the mind of man. *What*

---

* From the vantage point of congressmen from dirty air regions, however, uniform restrictions on new plants had additional attractions. NSPS meant that their states would be at less of a disadvantage in competing for new industry. *See* H.R. Rep. No. 1146, 81st Cong., 2d Sess (1970), *reprinted in* [1970] U.S. Code Cong. & Ad. News 5356, 5358; [1977] U.S. Code Cong. & Ad. News 1263 nn.2 and 4 (citing relevant sources); *cf.* Rose-Ackerman, *Does Federalism Matter?*, 89 J. Political Econ. (forthcoming 1981) (discussion of strategic implications of political behavior in federal system).

matters is the impact of sulfur oxide on the world outside the power plant. And this depends not only on emissions but on a host of other factors—the height of the stack, the direction of the wind, the plant's proximity to population centers. Unless EPA defined the "best system of emission reduction" to take into account these complexities, Section 111 might authorize a narrow inquiry into the technological design of the plant rather than a canvass of the ecological stakes involved in new construction. By giving statutory prominence to technological means of purification in new plants, Section 111 would distort policymaking perceptions for years to come.

Here we begin to glimpse the complexity of the link between environmental symbol and administrative process. However ironic, it is the New Deal ideal that is consistent with the ecological insight into reality. For it is the expert agency, unencumbered by abstract legalisms, that promises to craft a policy responsive to the complexities of environmental relationships. Yet the one thing clear to the Clean Air Act draftsmen was that the New Deal agency had failed. Moreover, their momentary success in engraving into the statute a demand for the "best" did not lead environmentalists to doubt their rejection of New Deal models of administration. The law had finally cut through bewildering complexities to promise our children a new world in which spanking new plants would churn out the old consumer goods in harmony with nature. Unfortunately, the law had also created a decision-making structure in which it might seem sensible for decision makers to define the "best system of emission reduction" in an ecological and economic vacuum—without a sober effort to define the costs and benefits of designing one or another technology into the plants of the future.

2

THE MYTH OF EXPERTISE

FROM STATUTE TO POLICY

In establishing its special regime for new plants, Congress did not try to take into account the peculiar problems of any particular industry. Instead, it was taking a step that would assure future generations that their interests would not be compromised by New Deal passivity. Despite Congress's aggressive stand on NSPS, it remained for the EPA to translate the statute into practical policy. Its task was to establish a standard that, in the words of the statute, "reflects the degree of emission limitation achievable through the application of the best system of emission reduction which (taking into account the cost of achieving such reduction) the Administrator determines has been adequately demonstrated."[1] As we shall show, however, setting such a standard would have been a tricky business at best—for in its desire to avoid the perils of expertise, Congress had adopted a simplistic formula that was exceedingly difficult to apply to the problem posed by coal burning.

Unfortunately, the EPA did not respond to the inept statute with creative use of its expertise, but treated NSPS as if it were merely a problem in applied sanitary engineering. Worse yet, when the issue came before the courts, the judges did not force the agency to adopt a sophisticated conception of its planning function. To the contrary, the press of environmental litigation only served to distort further the way policymakers perceived the ecological and economic stakes involved in controlling new plants. By 1975, issues were framed in ways that would have unhappy consequences in later years—when the status of new coal burners would become increasingly important to decision makers outside the EPA on Capitol Hill and in the White House.

Technology in a Vacuum

By its very terms, two aspects of the statute invited, if they did not require, agency sophistication in setting a standard for new coal burners. First, there was the instruction that the administrator take cost "into account" in making his decision. A billion dollars is a lot of money, but it is a sensible investment if it purchases ten billion in benefits. Rather than contemplate cost figures in the abstract, does the language not authorize an assessment of the *net* costs generated by a system *after* its environmental benefits have been taken into consideration?[2] Second, a proposed system had to be "adequately demonstrated" before it became the basis for a clean-up requirement. This phrase obscured difficult policy dilemmas. On the one hand, EPA could rely on new and untried clean-up technologies and take the risk that they would fail to work effectively. On the other hand, the agency could satisfy itself with well-established methods, and thereby lessen incentives to create new technologies that would improve clean-up performance. Yet there were creative possibilities for statutory interpretation that would ameliorate, if not eliminate, this tension. For example, could the agency not phase in a set of increasingly stringent requirements over an extended period of twenty or thirty years—thereby gaining certainty in the short run while allowing lead time for long-term innovation?

The agency did not try to ask, let alone answer, either of these questions. Instead, it read the statutory language as if a standard could be established on the basis of a narrow engineering judgment. To understand the bureaucracy's view of its own choices, we must consider the state of technology in the early 1970s. At that time, there existed two methods for reducing emissions from coal-fired plants: a relatively old-fashioned technique known as physical coal cleaning, or "coal washing," and an embryonic technology, flue gas desulfurization, commonly known as "scrubbing."

Physical cleaning removes sulfur from coal before the coal is burned. The simplest method involves equipment not much more advanced than a wire screen and garden hose: freshly mined coal is crushed, passed through a screen, and wetted, so that heavy sulfur-bearing fragments can settle out. These relatively simple processes can remove most of the sulfur-bearing particles, called pyrites, that are physically mixed with the coal as it comes from the mine. They cannot, however, remove sulfur that is chemically bonded to the coal. Nonetheless, the gains achieved by primitive washing techniques can be substantial, varying from 20 percent to 40 percent. Within these limits, physical coal cleaning is a cheap and reliable technology. Moreover, it was not obvious that potential gains from coal washing had been exhausted. By grinding coal into a fine slurry, it would be possible to wash more of the pyritic sulfur than could be reached by the primitive crush-and-hose methods practiced in the early 1970s.[3]

Nonetheless, the EPA disregarded such humdrum possibilities and concerned itself exclusively with more symbolically satisfying technologies—devices that, attached to a smokestack and paid for directly by the polluter, promised to cleanse the smoke produced by the boilers below. Among the technologies existing in 1971, the scrubber was the only one that performed this symbolic function. Its ability to perform as an acutal control technology, though, was more problematic.[4]

A scrubber does not rely on physical processes such as crushing and washing, but on the maintenance of a large-scale chemical reaction. It is a 70-foot test tube which on a typical day may consume 400 tons of limestone and thousands of gallons of water to remove over 200 tons of SO_2.[5] As exhaust gases flow up a power plant smokestack, they are exposed to a lime or limestone solution that is sprayed in their path. Sulfur dioxide in the gas reacts with the spray and goes into solution, from which it is later removed, dewatered, and extruded in the form of sludge. Maintaining the proper conditions for this reaction requires continuous supervision. For example, the coal burned may contain elements such as chlorine that interrupt the desired reaction; or a variation in the amount of SO_2 in the flue gas may adversely affect the process. Even when the reaction is proceeding apace, the machines must operate in a very harsh environment. Unreacted SO_2 and water may combine to form sulfuric acid which corrodes the inside of the scrubber and smokestack. Reacted SO_2 may form compounds that travel beyond the scrubber and clog smokestacks, pipes, and pumps. While these problems were not difficult to solve under carefully controlled conditions, the early scrubbers were prone to frequent breakdown—operating less than half the time.[6] When the administrator promulgated his new standard of performance in 1971, only three scrubbers were operating in the United States. The oldest, built in 1968, would be abandoned by the end of the year.[7]

In short, the EPA squarely confronted the problem posed by any statute requiring that a system be "adequately demonstrated"—how to trade off certainty and economy against incentives for further technological development. Moreover, creative ways of mediating the tension were available; perhaps a standard based on improved washing should have been required in the intermediate term with reliance on scrubbers projected for a decade or more hence. Rather than give the statutory formula a creative interpretation, however, the language was used as an excuse for thought. The possibility of advanced coal washing was totally excluded and official documents narrowly focused their attention on the question of whether the scrubber was "available" in some engineering sense, divorced from other development opportunities.[8] Such an approach justified the bureaucracy's spending its limited resources on engineering projections to determine if scrubbers could be made operational in the near future. On the basis of these projections, the administrator found that the scrubber's ability to eliminate about 70 percent of a coal burner's sulfur oxides had been "adequately demonstrated,"[9] and proceeded to the task of translating this engineering judgment into regulatory policy.

Squaring the Circle

At this point the agency was forced to confront the dilemmas imposed by its own impoverished reading of the statute. The embarrassing fact is that the agency's interpretation made it conceptually impossible to move from its engineering judgment about the scrubber's availability to a policy judgment defining the number of pounds of SO_2 a plant could emit for each MBTU of energy it produced.[10] To see why, consider that a power plant's emissions are not exclusively determined by its treatment technology but are also a function of the amount of sulfur in the coal that the plant burns. The agency's engineering judgment about scrubbing could readily be translated into an emission limitation only if *all* of America's coal contained the same sulfur content. Only then could 70 percent be multiplied by a constant to yield a single nationwide limit on new plant discharges. But, alas, America's coal reserves range in sulfur content from 1 to more than 10 pounds.[11] To make matters even more difficult for the administrator, these coals are distributed unevenly throughout the coal-producing regions. Roughly half of the nation's reserves lie west of the Mississippi in the Northern Great Plains and Mountain regions and consist largely of low-sulfur coal.[12] Eastern reserves, primarily from the Midwest and the Appalachians, contain much larger proportions of higher sulfur coal.[13]

These facts signaled the bankruptcy of the engineering approach: it is mathematically impossible to multiply a constant by a variable to yield a single nationwide numerical ceiling on power plant emissions. The agency's predicament was intensified by a final statutory artifact. Although the statute directed the administrator to look at the "best system" in defining applicable effluent standards, it did not authorize him to force polluters to install scrubbers if their power plants could meet the effluent limit in some other way.[14] Thus, whatever ceiling the administrator might set, polluters might find it cheaper simply to burn low-sulfur coal than to install scrubbers. The threat of a massive shift away from high-sulfur coal would, in turn, generate powerful political pressures from eastern producers. The conceptual inadequacy of the engineering approach—its lack of explicit concern with the variability of polluting inputs—masked a potentially explosive political problem: how to parry the predictable counterattack by eastern coal?

Surely not by surrendering without a fight. After all, the whole point of environmental regulation is to force producers to bear the social costs of their enterprise. High-sulfur coal had previously gained an unfair competitive advantage over low-sulfur coal precisely because the extra harm it caused the environment had not been reflected in coal prices. A program of controlling sulfur oxides merely removed that advantage.[15]

Although this point was obvious in the abstract, the eastern high-sulfur coal industry could hardly be expected to accept the loss of its advantage without a struggle. The EPA could have tried to preempt the foreseeable counteroffensive by showing that the threat to the industry was not nearly so great as a superficial analysis had suggested. Or it might have demonstrated that the dangers of sulfur oxides were so intolerable that the high-sulfur mines could not reasonably hope to defend their competitive advantage in the political struggle that was sure to follow. But in its initial 1971 decision, the agency did not prepare the ground for the battle that, even then, was plainly perceptible to knowledgeable observers.

Despite the conceptual impossibility and political unwisdom of moving immediately from its engineering judgment to regulatory standards, the agency tried to square the circle by treating the problem of coal variability as if it were a minor detail. The agency simply failed to recognize that its findings about scrubbing were compatible with an NSPS ceiling ranging all the way from 3 pounds per MBTU to 0.3 pounds per MBTU—depending on whether ten-pound eastern or one-pound western coal was being scrubbed at 70 percent efficiency. Apparently there was no effort, however rudimentary, to estimate the costs and benefits generated by a range of different possible emission ceilings.[16] Instead the agency finessed its conceptual and political problems by announcing a number and making a few casual remarks in its support. The numerical ceiling was set at 1.2 pounds of SO_2 per MBTU; in support, the EPA merely stated that this ceiling would permit eastern power plants to comply by scrubbing 70 percent of the SO_2 out of the average eastern coal—which was said to contain roughly 4 pounds of SO_2 per MBTU ($[1.0 - .70] \times 4$ pounds = 1.2 pounds).[17] At the same time, the agency recognized that utilities might respond to this ceiling the natural way, by burning 1.2-pound coal.[18] It failed, however, to estimate the impact such decisions would have on the eastern coal industry, let alone whether the benefits of the 1.2 standard outweighed its expected costs.[19] Instead, it blandly proclaimed that burning low-sulfur coal would also satisfy the new standard.[20]

In short, a study of the EPA's 1971 decision suggests that Congress had succeeded only too well in transcending New Deal ideals of expertise. Rather than interpret Section 111 as an invitation to put costs into the context of environmental benefits, the agency had fastened on the scrubber as a symbol of technological salvation. Rather than explain to high-sulfur coal producers why they would have to give way gradually before the public interest, the agency had tried to assume their problem out of existence.

Yet it was not as though the EPA had set out to make bad decisions. To the contrary, the agency had responded rationally to the bureaucratic incentives created by the Clean Air Act. In 1971, the EPA was an infant organization that had yet to create a reputation for itself; it confronted a congressional deadline requiring it to reach specified clean air targets by 1977 at the latest. If the EPA was to reach these targets, its prime problem was to force the states to force existing polluters to cut back on existing pollution, not to plan for plants coming on line in 1980 or 1990. To override the EPA's interest in establishing its credibility by acting in pursuit of short-term goals, Section 111 would have had to *require* the agency to begin the lengthy and difficult task of long-range planning. Yet this effort to force the agency to confront economic and ecological complexity was the last thing Congress had in mind in moving beyond the New Deal. Instead of forcing the agency to redeem the promise of expertise, Section 111 made it relatively easy for the EPA to treat NSPS as if it were a narrow engineering problem.

Moreover, the agency's narrow reading of the statute was not to be remedied by the creative use of judicial review. This is not to say that the District of Columbia Court of Appeals was altogether happy with the agency's performance. When the 1.2 standard was appealed by the utility industry, the court sensed something was wrong with the EPA's narrow conception of its mission. The particular fact that attracted judicial attention was the enormous quantity of sludge generated by a typical scrubbing operation. True to its mechanical reading of Section 111, the agency had failed to take this second-order environmental harm into account before finding that scrubbing qualified as the "best" method of emission reduction. This proved too much for the court of appeals, which remanded the matter to the EPA for explicit consideration. Unfortunately, the court did not recognize that the agency's disdain for the solid waste problem was only symptomatic of the abdication of its long-range planning function.[21] As a result, the agency merely repromulgated its original NSPS standard, this time explicitly declaring that it had taken sludge into account.[22]

Although the agency's continued neglect was readily explicable in terms of the bureaucratic needs of the moment, the sad story that followed owed a great deal to this bad beginning. The failure to create a sound structure for comparing costly means and environmental ends permitted partisans on both sides to pursue their interests without recognition of the larger aim—to fulfill our obligations to the next generation without causing unnecessary hardship for the Earth's present inhabitants.

THE DISTORTING PRISM OF LITIGATION

Yet we have not reached the end of judicial oversight during the early years when NSPS simmered on the back burner. While the EPA invested as little effort as possible on future planning, others were free to use the courts to try to force the agency to give NSPS a larger share of its scarce bureaucratic resources.

The Navaho Challenge

The principal lawsuit grew out of a conflict between chapters of the Navaho tribe and one of the environmental movement's perennial targets—the complex of massive coal-burning facilities located near Four Corners, Arizona.[23] Because low-sulfur coal was readily available in the Southwest, plants would be able to meet the 1.2-pound standard the natural way. In response, neighboring Navaho tribes argued that burning low-sulfur coal was not sufficient under Section 111. They insisted that the EPA require the plant to install scrubbers, thereby reducing SO_2 well below the 1.2 NSPS standard.[24]

From the vantage point of a litigator concentrating on Four Corners, this demand was but another battle in the long war to impose costly controls for environmental gains. From the national perspective, however, the issue was far more complicated. Southwestern power plants would continue to use western low-sulfur coal regardless of scrubbing requirements for the simple reason that, given the heavy expense of railroading high-sulfur coal from the East, nearby low-sulfur coal remained the

cheapest option. Yet, although a 70 percent scrubbing[25] requirement would reduce Four Corners' emissions by 70 percent, this
would hardly be the outcome of a nationwide requirement. In
the East and Midwest, many plants would have a choice between
nearby high-sulfur coal and more distant low-sulfur varieties. If
the ceiling remained at 1.2 and scrubbing were required by law,
these utilities would lose all economic incentive to pay the heavy
costs of shipping low-sulfur coal. Instead, they would use their
70 percent scrubbers on the four-pound coal readily available
nearby. Hence, a 70 percent scrubbing requirement need *not*
yield a 70 percent reduction in emissions. Instead, it might simply induce many utilities to substitute scrubbers for low-sulfur
coal in an effort to comply with the 1.2 standard. As far as the
East was concerned, forced scrubbing promised to be a symbolically satisfying but peculiarly inefficient response to the realities
of coal burning.

Worse yet, a massive shift to scrubbing would place enormous
burdens on the EPA inspection system.[26] Scrubbers only scrub
when operated; unless EPA mounted an ongoing and credible
enforcement effort, 70 percent scrubbing would be an un-
redeemed promise. If utilities did not in fact operate their
scrubbers at high efficiency, the shift from low- to high-sulfur
coal might even *increase* the SO_2 loadings east of the Mississippi.
The environmental issue, then, was squarely posed: putting
aside other costs, were the gains in the West worth the threat of
increased sulfur loadings in the East?

Given their narrow focus on Four Corners, however, the Navaho had no need to concern themselves with the impact of
scrubbing on the other side of the country. Worse yet, when the
Navaho went to court to force the EPA to impose scrubbers, the
agency did not try to defend the 1971 NSPS decision on its merits. Although there is evidence that agency specialists were aware
of the costs of inducing a shift away from low-sulfur coal in the
East,[27] agency lawyers chose to defend EPA's 1971 decision on a
narrow procedural ground. In this they were successful—the
District of Columbia Circuit Court rejected the Navaho chal-

lenge, relieving the agency of the immediate need to invest more analytic energy in the scrubber.[28] Unhappily, however, the lawyers' strategic victory had deprived the agency of a critical opportunity to educate its public. Although Four Corners dramatized the way scrubbing might reduce western emissions, no similar event would dramatize the dangers of a high-sulfur coal strategy in the East.

The Struggle over Existing Plants

While procedural victory kept NSPS on the back burner, all participants turned their immediate attention to the critical issue raised by the passage of the Clean Air Act. In 1970, hundreds of coal-burning power plants contributed about half of the overall load of sulfur oxide in the United States.[29] Rather than speculate about future plants, the obvious problem was posed by existing ones.[30] If the act's promise of clean air by 1977 was to be redeemed, many of these plants would be required to cut back their emissions substantially.[31] Moreover, under the law, each plant's lawful emissions could be determined only after a state agency had calculated the cutbacks required to assure local compliance with the ambient standards for sulfur dioxide.[32] A set of exhausting courtroom battles lay ahead.

Predictably, utilities tried to minimize the cost of the changes required of them. Instead of building scrubbers or burning low-sulfur coal, a coal burner could reduce its effective contribution to the local sulfur oxide problem by building tall smokestacks, up to 1,000 feet high, thereby exporting more of its emissions to distant regions.[33] To respond to the temporary higher concentrations created by inversions or other unfavorable meteorological conditions, the utilities proposed use of "intermittent control systems": either they would hold a reserve of low-sulfur fuels to burn during inversion periods or they would bring their cleanest plants on line first to meet peak demand conditions.[34]

The environmentalists responded to these proposals with alarm. Tall stacks, they rightly feared, would improve local conditions only at the expense of more distant areas. They also opposed the utilities' use of intermittent controls, which imposed significant burdens on future efforts to monitor compliance.[35] In response to these fears, environmental groups brought suit to challenge the legality of the utilities' strategies under the Clean Air Act.[36] Once again, however, the exigencies of litigation obscured the real problems raised by sulfur emissions. Although the EPA recognized that the long-range transport of sulfur oxides raised serious policy problems,[37] environmentalists did not concentrate on forcing EPA to channel its energies in long-term directions. Instead, they sought solutions more readily obtainable through judicial decree.[38] Moreover, the desire to attain rapid compliance with SO_2 standards produced a peculiar faith in technological solutions. Rather than press power plants to use low-sulfur fuel, environmentalists emphasized scrubbing as their preferred remedy to the utilities' foot-dragging.[39]

While implementation of any clean-up strategy awaited the outcome of litigation,[40] the rhetorical battle over scrubbing—both in and out of court—heated up. The utilities insisted that scrubbers were unreliable, and environmentalists increasingly viewed scrubbers as the final solution to the power plants' unreasonable delays.[41] Thus, the old-plant dispute evolved in a way that mocked the aspirations of the SIP process—to tailor clean-up burdens in a way that was sensitive to environmental realities. Instead of focusing on the long-range transport of sulfur oxides, partisans perceived the critical issue to be the viability of a single technology. And though the scrubber was initially considered in connection with old plants,[42] it would naturally come to mind as a "solution" to the new plant problem when it once again became a live policy issue.

LITIGATION AND LONG-RANGE PLANNING

There are many lessons to be learned from the first five years The most obvious is that we cannot rely on the courts to make up for the failures of the administrative process. Not only did the EPA readily evade the court of appeals' effort to prompt policy reconsideration, but the flow of litigation only directed attention away from the need for long-range planning. Nor was this distortion an unhappy accident. Environmental litigation is typically generated by actions in the here and now that catalyze environmental anxieties—the building of a new plant, the refusal to clean up an old one. No less important, a successful lawsuit against a particular plant will (ultimately) yield a palpable sense of victory for both the environmental lawyer and his clients. In contrast, a lawsuit to compel the agency to engage in sophisticated long-range planning is a daunting prospect. Not only is it harder to sustain public interest in a complex and esoteric debate, but there is a danger that the lawsuit will never end—with the lawyers sinking without a trace in an endless series of remands and reconsiderations. While environmental lawyers deserve great praise for taking on a surprising number of these unrewarding lawsuits,[43] the cases discussed here—and many others—suggest that litigation obscures as much as it instructs. There is no substitute for creating *bureaucratic* incentives that will reward officials themselves for informed decisions. After all, that is what bureaucrats are supposed to be paid for.

Yet it is important to maintain a balanced view of the early agency performance. While we have no doubt that the 1971 NSPS decision could have taken a far more sophisticated approach to the problem, the fact remains that it was made at a very early stage of the EPA's existence. And once the initial decision had been made, it was not silly for EPA policymakers to refuse to squander more of the agency's scarce resources on immediate reconsideration. Yet, as the coal burner rose in public importance, it was inevitable that a serious reappraisal would occur. How would our institutions make use of the opportunity?

3

THE POLITICS OF IGNORANCE

FILLING THE ANALYTIC GAP

When reconsideration came, it was not to be attempted in the old New Deal way—by agencies generating policy under the uncertain supervision of courts. Congress itself was to intervene significantly in the resolution of the scrubbing controversy. Legislative intervention was no happenstance; it was the inevitable consequence of the Clean Air Act's attempt to move beyond the New Deal. Operating under a vague statutory mandate, New Deal agencies have ample opportunity to fend off statutory interventions by adapting to changing congressional sentiment through the use of administrative discretion. In contrast, the Clean Air Act had tried to resolve so many disputable substantive issues in 1970 that recurrent congressional reconsideration was fundamental to the policymaking process.[1]

Nonetheless, the administrative history we have just recounted is of great importance; it provides a perspective for appraising Congress's institutional performance. As we have seen, the EPA had successfully avoided the need to develop a full-blown analysis of the costs and benefits of NSPS in general and of scrubbing in particular. Now that Congress was to intervene, would it fill this analytic gap and try to view the scrubber in economic and environmental perspective?

If the fate of the scrubber had been left to a New Deal agency, at least such an investigation would have been assured. While the analysis might well have contained errors, the agency would have risked judicial reversal if it had failed to make a plausible-looking effort to assess the nationwide costs and benefits of scrubbing. Yet even this minimal effort proved to be beyond Congress's institutional capacity. A bizarre coalition of environmentalists and dirty-coal producers tried to exploit congressional ignorance to serve their own, mutually incompatible, purposes. Congress was never invited to look beyond the Four Corners controversy to confront the economic waste and environmental dangers generated by forced scrubbing in the East. The result of congressional intervention was a hopelessly incoherent mix of statutory language and legislative history that set the stage for the most recent effort by the EPA to reconcile the law to the environment.

The congressional story will be in two parts. This chapter analyzes the emergence of forced scrubbing on the legislative agenda during 1975 and 1976. Chapter 4 discusses the remarkable way Congress dealt with the scrubber as part of a larger effort to legislate answers to a host of questions raised by its earlier attempt to move beyond the New Deal.

ASKING THE WRONG QUESTIONS

Within the statutory scheme as a whole, the critical year was 1977. That was when two major objectives were to be achieved by congressional decree. First, the auto industry was to achieve stringent emission cutbacks initially established by the Congress of 1970.[2] Second, 1977 was the year all America was finally to enjoy a level of air quality sufficient to "protect the public health" with "an adequate margin of safety."[3] As early as 1975, however, it was plain that these objectives would not be met and that further congressional action would be required in the years ahead. To establish the foundation for its future interventions, both House and Senate subcommittees began serious preparatory work.

As in all such efforts, the initial work did not fall to Congress as a whole, or even to the congressmen on the relevant subcommittees, but to congressional staffers. Moreover, these people did not imagine that they were to write clean air amendments on a clean slate. Instead, they permitted the controversies of the act's early years to define the questions, not to mention the solutions, that were worth worrying about.

The scrubbing issue surfaced directly in the 1974–75 hearings before the House Subcommittee on Health and the Environment. The discussion at the hearings revealed the pervasive influence of the ongoing course of litigation. Thus the scrubbing question was almost exclusively discussed in terms of old plants rather than new ones.[4] It was the continuing failure of these plants that jeopardized the successful attainment of SO_2 objectives by 1977. Moreover, the environmentalists could easily portray the utilities' response—litigation plus a *promise* of "intermittent controls"—as an obvious attempt to exploit the act's weaknesses in order to avoid a costly clean-up effort.[5] In contrast, the status of new plants was a low-visibility concern. While House staffers were aware of the Four Corners controversy, neither EPA nor anybody else informed them of the complex environmental issues that Four Corners concealed.[6]

To the extent that the future coal burner received focused attention, it was in connection with a second issue that had first been defined by litigation. While the 1970 act plainly established minimum air quality objectives for the nation, many areas already enjoyed air far cleaner than the act required. Yet the original draftsmen had not squarely confronted the problem posed by this embarrassment of riches. In these areas, would the 1970 act amount to a Dirty Air Act, legitimating pollution up to the minimum standards? Congress's failure to ask this question generated a characteristic effort by agencies and courts to fill the gap. When the EPA began to approve SIPs that permitted power plants to degrade clean air regions, the Sierra Club went to court to challenge the legality of this decision. It emerged victorious by a narrow margin. After it had convinced a single district judge of the impropriety of the EPA's decision, the outcome was sustained by an equally divided Supreme Court.[7]

Unless Congress amended the statute, this decision meant that EPA could not approve state plans which permitted increases in the levels of pollutants in the clean air areas. Although environmentalists strongly supported this result, their objectives would evoke concerted opposition by groups interested in western economic development.[8] In search of ways to package environmentalist goals in politically acceptable forms, House staffers turned to the solution advanced by the Navaho's abortive lawsuit at Four Corners—a scrubbing requirement for all new plants. Even if the highly visible subcommittee proposals mandating "nondegradation" of the clean air areas succumbed to political counterattack, the subcommittee might provide backup protection to the clean air of the West by requiring all new coal burners to use scrubbers.

THE HOUSE PROPOSALS OF 1976

In short, universal scrubbing would provide a second line of statutory defense for the clean air regions. When amendments were initially proposed by the subcommittee, they contained not only elaborate provisions to prevent the significant deterioration of clean air regions (or PSD,* as it is known in the trade) but also a barely perceptible alteration of Section 111. The 1970 version of this section directed the agency to set performance standards that reflected "the degree of emission limitation achievable through the application of the best system of emission reduction. . . ."[9] In contrast, the new House proposal required a standard which "reflects the degree of emission *reduction* achievable through the application of the best *technological* system of *continuous* emission reduction . . ." (emphasis added).[10]

The significance of this change was far from obvious, given a

* For Prevention of Significant Deterioration.

special definition of "technological system of continuous emission reduction" that also appeared in the House proposals. Rather than define the term to require an "add-on" pollution control technology, Section 111(a)(7) gave the term a technical meaning to include "a technological process for production or operation by any source which is inherently low-polluting or nonpolluting."[11] The use of low-sulfur coal seems precisely the kind of step suggested by this explicit definitional extension —after all, a shift from high- to low-sulfur coal can reduce power plant emissions by 80–90 percent! If anything, this wording placed the use of low-sulfur coal on a more solid statutory foundation.

Turning to the House committee report, however, one enters a new world of meaning. It proclaims that scrubbing or some other add-on technology is required of all new coal burners. The specific definition of "technology" to include the operation of "low-polluting or nonpolluting sources" is explicitly denied— without explanation—to the typical user of low-sulfur coal.[12] This clear statement would bear fruit only if the EPA and the courts focused their attention upon the committee report, glancing at the statute itself only when the report's language was ambiguous. Unfortunately, however, this approach to statutory interpretation is hardly unprecedented.[13] If the legislative history were to be taken seriously by the courts, the equation of a "technological system" with scrubbing would serve the subcommittee's purpose of providing low-visibility insurance for the clean air areas.

Despite this threat to their interests, however, the utility industry was more concerned with a more visible statutory development. Rather than focusing upon the legislative history being manufactured under Section 111, the utility lobby concentrated its forces on the clear and present danger presented by the committee's explicit statutory effort to protect "clean-air regions" through its PSD mechanism.[14] While the utility industry marched off to battle against PSD, House staffers began to organize political support for the invisible amendment of Section 111.

Their search for politically potent allies for the "new" NSPS led in a surprising—but readily explicable—direction. Although universal scrubbing had only a problematic relation to clean air goals, there *was* an interest group that valued it for its own sake—the producers of high-sulfur coal in the eastern United States. For many years this group had fought and lost the good fight against "unnecessarily" stringent pollution standards.[15] As far as they were concerned, the invisible amendment of Section 111 offered a new strategy that promised to further their basic interests more effectively. Once the eastern utilities were forced to install scrubbers, it would be possible for them to meet the 1.2 NSPS while continuing to use cheap high-sulfur coal. Only if utilities were allowed to substitute low-sulfur coal for scrubbers would a shift away from high-sulfur products be conceivable. Thus it made sense for the dirty coal producers to abandon their campaign to weaken pollution standards and take up the cudgels for the costliest possible clean air solution—universal scrubbing.

As a consequence, when House staffers made contact with eastern coal interests, they met with a sympathetic response. From the point of view of the United Mine Workers Union, the scrubbing issue was particularly straightforward. Because its membership is concentrated in the East, it had no difficulty coming out publicly for universal scrubbing.[16] Politics were more complicated for the National Coal Association. Since western owners were naturally interested in maximizing sales of low-sulfur coal in the East, they would not take kindly to the national lobby endorsing a requirement that would freeze them out of a potentially rich market. Consequently, the National Coal Association refused to take a position on scrubbing, leaving it to eastern members to use their considerable political muscle in ways we shall describe.[17]

The stage had been set for a bizarre coalition between clean air and dirty coal forces. The House subcommittee had originally turned to scrubbing as an ancillary environmentalist measure in support of PSD and viewed the coal lobby as a convenient ally in the battle for clean air. But once the attention of the coal lobby had been engaged, the scrubbing issue took on a life of its own in the service of regional protectionism. The consummation of this political marriage was evident as early as the 1976 House committee report, whose language was repeated in the House report the following year. Besides announcing that the "new" Section 111 had invisibly imposed a universal scrubbing requirement, the report justified the innovation by finding six flaws in the "old" NSPS:[18]

1. The standards give a competitive advantage to those States with cheaper low-sulfur coal and create a disadvantage for Mid-western and Eastern states where predominantly higher sulfur coals are available;

2. These standards do not provide for maximum practicable emission reduction using locally available fuels, and therefore do not maximize potential for long-term growth;

3. These standards do not help to expand the energy resources (that is, higher sulfur coal) that could be burned in compliance with emission limits as intended.

4. These standards aggravate compliance problems for existing coal-burning stationary sources which cannot retrofit and which must compete with larger, new sources for low-sulfur coal.

5. These standards increase the risk of early plant shutdowns by existing plants (for the reasons stated above), with greater risk of unemployment.

6. These standards operate as a disincentive to the improvement of technology for new sources, since untreated fuels could be burned instead of using such new, more effective technology.

Although this pronunciamento was written by House staffers with strong environmentalist reputations, its authors might easily have come from a major coal company. While point 6 attempts an inadequate invocation of technology-forcing,[19] the text employs the standard rhetoric of the eastern coal lobby—the need to eliminate unemployment by using "locally available fuels" and to defeat the energy crisis by burning high-sulfur coal. The committee report, moreover, contains no data to support this rhetoric—at a time when readily available data would have placed these pleas of the eastern coal lobby in a very different light.

THE USES OF TECHNOCRATIC INTELLIGENCE

While political muscle would be required to induce Congress to pronounce upon the scrubber's desirability, something more was also needed: an appropriate technocratic show. As the House committee position evolved in 1975 and early 1976, the committee staff turned to the executive branch to provide data in support of its proposals. In response, a group of air quality modelers from the EPA and the Federal Energy Administration (FEA) prepared a set of staff studies.[20] The technocrats' principal task was to examine the impact of the highly controversial PSD proposals. But the EPA staff also cast a sidelong glance at the House efforts to impose universal scrubbing through Section 111.[21] Instead of using their expertise to redefine the problem as it was perceived by busy laymen on Capitol Hill, the EPA staff's work proceeded within the framework of congressional understanding. It failed to address the environmental risks of scrubbing high-sulfur coal in the East.[22] Even more remarkable, the EPA did not ask itself whether there were better ways of reducing SO_2 emissions than the universal scrubbing policy that seemed to have gained House favor. Most important, it failed to assess the possibility of achieving emission reductions more cheaply by lowering the 1.2 ceiling rather than requiring all coal burners to use scrubbers. Thus it presented a false dichotomy: universal scrubbing was treated as if it were the *only* alternative to EPA's existing 1.2 standard. Because cheaper ways of achieving lower emission levels had been excluded a priori, the formal analysis made it appear that Congress was buying some real sulfur oxide emission reductions by forcing scrubbing on the utility industry.[23]

Despite its narrow focus, the technocratic report was hardly an unqualified endorsement of the subcommittee's innovation. The EPA emphasized that the invisible amendment to Section 111 was a very expensive proposition—by 1990 adding 14 billion dollars over the level of expenditure required by PSD alone.[24] Moreover, the data provided on eastern coal did not support the prominent place given protectionism in the committee report's rhetoric. While a western mining boom was predicted by 1990, this gain did not imply a dramatic destruction of the mines east of the Mississippi. Even without forced scrubbing, eastern production would increase from 531 million tons in 1974 to 680 million tons in 1990.[25] To refine its analysis of market impact, the EPA took account of the fact that eastern production is concentrated in two distinct geographic regions. In Appalachia, the study predicted growth of 45 percent by 1990. The model's optimism was justified by two factors. First, the region contains very substantial quantities of low-sulfur coal, particularly in southern West Virginia and eastern Kentucky. Second, transportation costs are very important in the coal business—at a certain point, it makes economic sense for a utility to install costly scrubbers rather than pay the price of shipping low-sulfur coal from distant regions. In contrast to the happy picture in Appalachia, the model depicted a darker future for the second eastern area—centered in Illinois, Indiana, and western Kentucky. Given the high-sulfur content of midwestern coal and its relative proximity to the low-sulfur West, the EPA predicted that coal production would decline by 15 percent from 149 to 125 million tons by 1990. Here universal scrubbing would make a perceptible difference—the EPA predicted that, if forced to scrub, midwestern utilities would buy an extra 25 million tons of local coal in 1990, almost precisely offsetting the predicted regional reduction. Considering the East as a whole, however, this 25 million ton gain was less than 4 percent of actual production and

represented no more than 14 percent of the total growth projected for the region under full scrubbing. Given the 14 billion
dollar cost to the nation's consumers, the price of coal company
rhetoric would be high. Perhaps that is why this aspect of the
EPA staff's work is not mentioned in the House committee's majority report.

While it is easy to see why the House subcommittee would ignore the disturbing implications of the EPA staff's work, the
predictions do place a new perspective on the peculiar political
coalition that was forming on behalf of universal scrubbing.
After all, if the EPA's figures were close to being right, the en-
thusiasm of eastern coal producers for scrubbing is hard to understand. Their motivation becomes clear, however, if the anxi-
ety of eastern coal is explained in terms of its recent past rather
than its foreseeable future. For there *was* a time eastern coal had
plainly suffered at the hands of clean air. Paradoxically, this
time largely antedated the passage of the Clean Air Act of 1970.
The problem had its source in the 1960s when northeastern
states dramatically cut back on the sulfur emissions they considered tolerable. These state initiatives induced northeastern utilities to substitute imported low-sulfur oil for eastern high-sulfur
coal. Coal shipments to the Atlantic coast fell from 11 million
tons (or 70 percent of all utility fuel burned) in 1964 to 5.9 million tons (or 15 percent of all fuels) in 1973.[26] Although this col-
lapse in the market was offset by gains in coal sales in other
areas,[27] the coal producers had been taught a lesson they would
not quickly forget. This historical reality, not the comforting
computer printout, most impressed coal interests when they responded to the long-run threat posed by the gradual replacement of old plants with new ones.

POLITICAL CONVENIENCE AND LEGAL AMBIGUITY

While the House report established the foundation for a clean
air–dirty coal alliance, it remained for the environmental lobby

in Washington, D.C. to accept the political invitation. Looking at the matter from New Haven, Connecticut, this acceptance does not seem preordained. Like their utility industry rivals, the attention of leading public interest lawyers was focused upon the PSD controversy and the treatment of existing plants.[28] Although these controversies predisposed them favorably toward scrubbing, they were also aware of the heavy cost of imposing particular technologies on industry rather than permitting firms to search out the cheapest ways of meeting emission requirements.[29] They also had a great deal of practice ferreting out assumptions buried in technocratic reports—it would not have been hard to flush out the EPA's false dichotomy between universal scrubbing and the preexisting 1.2 standard, which ignored cheaper alternatives for reducing emissions.

Instead of campaigning for a congressionally mandated reduction in the 1.2 standard, however, public interest lawyers embraced the dirty coal rhetoric. For example, Joseph Brecher, on behalf of the Sierra Club, condemned the 1971 NSPS decision because "eastern high-sulfur coal, which is now available, is having a hard time getting a market because of the comparative cheapness of bringing in western low-sulfur coal."[30] Richard Ayres, for the Natural Resources Defense Council, bemoaned the fact "that the Clean Air Act became a factor influencing the competition in the marketing for coal, encouraging practices such as shipping mountain state coal to Illinois and Louisiana, with attendant use of oil powering the diesel engines* used to transport train after train of coal."[31] These passages represent a remarkable rhetorical turn. Typically, environmentalists do not protest when a government initiative forces industry to discard "dirty" inputs and substitute "clean" ones. In so doing, government is simply eliminating market distortions and requiring the

---

* In contrast to the suggestion made by Ayres, it is now predicted that the free use of low-sulfur coal, instead of the mandatory use of scrubbers, will decrease the nation's dependence on oil. *See* note 9, chap. 6, *infra*.

prices of high-polluting products to reflect their true social cost. Rather than condemn the advantage gained by clean coal as artificial, environmentalists characteristically applaud when the "true" costs of dirty coal have finally been revealed.[32] No matter. With such rhetorical assistance, the peculiar coalition between the friends of clean air and dirty coal would be a powerful political force.

Yet rhetoric alone cannot overcome the fundamental conflicts concealed by a marriage of convenience. While both groups wanted scrubbers, they wanted them for different reasons. Environmentalists saw scrubbing as a technique for cutting new plant emissions *below* 1.2 pounds to even lower levels.[33] For example, if a plant using one pound western coal were required to scrub at 90 percent efficiency, only 0.1 pound of SO_2 would be discharged. For eastern coal interests, however, scrubbing was desirable only as long as new plants could keep discharging at the old 1.2 level.[34] If the administrator used scrubbing as a reason for lowering the emission standard dramatically below 1.2, high-sulfur coal producers would be frozen out of the new plant market once again.

We return to the master policy problem that the EPA finessed in 1971 when it first established new source standards for coal burners. Given sulfur variability in coal, it is conceptually impossible to move unproblematically from the availability of scrubbing to the acceptable amount of SO_2 coming out of the smokestack. Instead of focusing on technological means of purification, the critical question was whether the existing 1.2 ceiling was too high, too low, or about right. In 1971 the administrator had tried to answer this question in a low-visibility, question-begging, way. Now that the issue was rising in prominence, the peculiar convergence of clean air and dirty coal created a political situation in which the combine would try its best to avoid confronting the policy problem posed by sulfur variations in coal. Moreover, the House staff's drafting strategy made it easy for Congress to avoid a clear policy directive. By avoiding an explicit statutory amendment and manipulating legislative history, the House subcommittee promised both political supporters the best of both worlds. While the House report clearly asserted that the administrator could remain content with the 1.2 emission limit—thereby satisfying eastern coal—the fact that this assertion was "merely" contained in the legislative history meant that the administrator remained free to use the new technology as a reason for cutting back on the emission ceiling—thereby satisfying the environmentalists. Rather than provide statutory guidance, the House left this key policy issue in precisely the same condition it found it. Only this time, when the administrator was confronted with the decision, he would no longer be able to evade the problem as he had in 1971. Instead, he would confront parties who thought that their political victories in the committee reports had entitled them to legal victories in the administrative process.

LOW-VISIBILITY POLITICS

As far as the partisans of scrubbing were concerned, no good would come from enlarging the debate on the premises of the new Section 111. Public discussion could only reveal the disparate hopes that bound the clean air–dirty coal coalition together. Moreover, the technocratic support work failed to signal the existence of a distinctive policy problem. Although the EPA study revealed the high costs of universal scrubbing, it also created the illusion that these costs were necessary to achieve further reductions in sulfur dioxide emissions. At only one point in the committee's work was there any warning sign that universal scrubbing would not necessarily lead to cleaner air. After plowing his way past 408 pages consisting of the majority report and appended supplements, the determined reader would find another 100 pages in which various congressmen expressed a variety of opinions on the myriad issues raised by the amendments. The only sustained discussion of NSPS appears on two pages of a thirty-page review by eight, mostly conservative, Congressmen:[35]

> Under current law (as defined by EPA) these sources would be allowed to choose the most cost-efficient means of compliance (for example, low-sulfur coal or scrubbers). However, under section 111 of H.R. 6161, which amends the current definition of standard of performance, new major sources in many instances must employ an extremely costly device known as a scrubber, even if the ambient air quality standards can be both attained and maintained by the use of low-sulfur fuels alone and even if the major source is located at a site producing low-sulfur coal.
>
> We can see absolutely no meaningful justification for such a costly and wasteful policy. We should pursue mechanisms which will assure that the ambient air quality standards are fully protective of our health and welfare and we must actively enforce the implementation plans to further assure that such standards will be fully attained and maintained. However, we should avoid the policies which will impose tremendous cost burdens on our electric utilities and on the citizens of this Nation when such costs are unnecessary for the attainment and maintenance of the national ambient air quality standards.

This effort to separate NSPS from related issues had no significant impact on the course of congressional debate. Because the utility industry concentrated its assault on the very idea that Congress ought to give special protection to clean air areas, no significant lobby invested resources in documenting the shaky relationship between universal scrubbing and *any* of the aims that Congress was considering. After the new Section 111 left the House committee, nothing was said that suggested general congressional awareness of the unnecessary expense, and doubtful environmental benefit, generated by forced scrubbing.

In part, failures like this are an inevitable consequence of the decision by the 1970 draftsmen to move beyond the New Deal. Congressional attention is one of the scarcest resources in the political system, and the administrative agency is one of our principal means of economizing on this resource. Some of the time, however, the executive branch will compensate for lapses in congressional awareness and focus attention on "sleepers" that would otherwise be enacted through collective inadvertence. Moreover, as the act moved through the Congress, the Ford administration did cast a skeptical eye upon Section 111. Using the same technocratic work as the House committee, the administration recognized that PSD in no way required universal scrubbing in all areas of the country. Impressed with the 14 billion dollars in estimated extra cost, the Ford administration proposed a compromise under which a universal scrubbing requirement would be postponed until 1985—by which time Congress would be obliged to revise once again its increasingly complex legislative creation.[36] As with the House committee, however, the administration saw Section 111 as ancillary to the PSD issue. The effort to delay universal scrubbing was part of a compromise in which the administration would, for the first time, announce its acceptance of PSD in principle. When this concession was bitterly attacked by Senate Republicans and industrial groups, the administration withdrew its compromise —failing to distinguish clean air objectives like PSD from Section 111's low-visibility effort to require scrubbing *regardless* of its contributions to any clean air objective.[37]

Given its failure to recognize the distinctive problem raised by universal scrubbing, the administration's withdrawn compromise contributed nothing to the evolution of the act. The invisible revision of Section 111 passed through the House untouched; its lack of salience was emphasized by the failure of the Senate committee to include a comparable revision in its legislative proposals.[38] When a House-Senate conference committee reported its compromise set of amendments, however, the House revision of Section 111—with its legislative history—was included without further elaboration.[39]

At this point, the conference compromise ran into the shoals of industry opposition. After studying the text, the auto industry decided that it could do better for itself by delaying final action until the next Congress—when all of Detroit's new car production would be banned under the Clean Air Act unless "realistic" amendments were forthcoming. At the same time, western Congressmen were encouraged by regional business interests to view PSD with alarm. Armed with the maps of areas doomed to economic stagnation in the name of supercleanliness, western industry prepared for a maximal effort against PSD. This combination of western industry and midwestern autos proved sufficient to block the act. When Senator Jake Garn of Utah began a filibuster, the conference bill was put to one side with remarkable speed—as congressmen prepared for the 1976 election.[40]

4

THE MYTH OF MAJORITY RULE

ENERGY POLICY TAKES CENTER STAGE

On January 20, 1977, a new Congress and a new president confronted some old problems: the Clean Air Act still threatened to
halt all industrial growth in the large areas of the country that
had failed to heed the congressional demand for clean air by
1977; similarly an "intransigent" auto industry was threatened
with shutdown as a result of its failure to reach the 90 percent
emission cutback mandated in 1970.

The political problem was further complicated by the new administration's desire to begin its four years on the upbeat. As far
as President Carter was concerned, the "energy crisis" had replaced the "environmental crisis" as the central problem on his
domestic agenda. As a consequence, the new administration did
not give first priority to the recurring air act controversy, contenting itself with a low-level bureaucratic response. When EPA
presented its "new" position on the Clean Air Act's revision, it
handled the low-visibility NSPS issue in a way that revealed the
years the problem had spent on the agency's back burners.
Rather than develop the environmental difficulties it had
glimpsed when resisting the Navahos' attack on Four Corners,
EPA simply regurgitated the 1976 House committee report's
dirty coal rhetoric, emphasizing the economic importance of
burning locally available coal.[1]

In contrast, the president committed the executive's resources
to James Schlesinger and assigned him the humbling task of
devising a comprehensive energy plan in ninety days. In this

context, scrubbing once again appeared an ancillary concern. Independence from foreign oil could be achieved only by burning more domestic coal. Yet to make coal burning politically acceptable, Schlesinger would have to come up with something that would calm environmental anxieties. It was this logic that led quickly to forced scrubbing.

To meet the president's deadline, Schlesinger cut off his staff of technocrats from the interminable process of bureaucratic consensus building.[2] Thus he was not in a position to explore the complex questions of environmental policy raised by scrubbing—moreover, even if he had consulted EPA, the agency was hardly in a position to present a concrete analysis of the impact of scrubbing high-sulfur coal on the health of easterners. Nevertheless, despite his isolation and the press of other issues, Schlesinger received—within four weeks—a far more cogent analysis of scrubbing than the one presented by the EPA in its formal comments to the new Congress on the Clean Air Act. While the briefing paper did not explore the environmental complexities, it did bring a few basic points to the forefront. First, it emphasized that a concern with clean air might be better expressed by reducing the present 1.2 emission limit to some lower figure. Second, it revealed the regional economic conflict concealed by the environmentalist rhetoric—stating that universal scrubbing "could slow down coal development in the West and encourage development of locally available coal." Third, it estimated that scrubbing would "raise utility capital requirements by 3–4 percent and electricity bills by 2–3 percent. Regional impacts could be higher."[3]

While such costs—which amounted to billions of dollars—would merit serious study in other arenas, they seemed minor when viewed from the Olympian heights of comprehensive energy planning. With hundreds of billions of dollars at stake with the stroke of a pen, scrubbing was pretty small change. Moreover, the political benefits of a simple solution to the environmental problem seemed very great indeed. The abortive effort by Nelson Rockefeller to push a comprehensive energy plan through Congress during the previous administration only emphasized that a long and difficult journey lay ahead. There would be enough battles to fight without taking on ones that could be avoided. The environmental lobby had proved a congressional power, and scrubbing was an easy way of converting a dangerous opponent into a formidable ally. Despite the technocratic critique, Schlesinger attached a forced scrubbing requirement as an environmental safeguard on the energy plan's efforts to encourage utilities to convert from oil to coal.[4]

With scrubbing now part of Schlesinger's comprehensive energy plan, it was no longer appropriate for the EPA, or anyone else in the executive branch, to ask hard questions about the relationship of scrubbing to any of the *environmental* goals before Congress in the 1977 clean air legislation. To the contrary, a carefully coordinated lobbying effort—featuring, among others, Douglas Costle, the administrator of the EPA—stressed the ways the president's plan was compatible with legislation already pending before Congress.[5] As a consequence, the House subcommittee had no incentive to reconsider its change in Section 111, and the coal lobby rhetoric of the committee report remained essentially unchanged. Moreover, the new controls over new power plants were not the subject of a special vote in the House but passed as an undifferentiated part of the legislative package.[6]

REGIONAL PROTECTIONISM IN THE OPEN

While the scrubbing proposal hidden beneath Section 111 languished on the periphery of legislative attention, a new initiative for the first time forced Congress to focus on the peculiar political logic behind forced scrubbing. The issue of regional protectionism was raised by the act's treatment of old plants rather than new ones. After the defeat of a frivolous amendment,[7] Congressman Rogers, the bill's floor manager, proposed a new floor amendment that displayed the same political logic that had inspired his committee's revision of Section 111. The proposal, which evolved into Section 125 of the act, was entitled: Measures to Prevent Economic Disruption or Unemployment.[8] As finally enacted, it permitted any governor, the EPA administrator, or the president to find that a shift from "locally or regionally available coal" would result in "significant local or regional economic disruption or unemployment." On such a finding, either a governor or the president could order a power plant to use regional coal to meet the applicable emission requirement. Such an order, of course, would mean requiring a coal burner to install scrubbers to bail out local miners and mine owners. The amendment, which gained the explicit endorsement of the United Mine Workers, represented an enormous step in the evolution of the Clean Air Act. No longer was the act's symbolic demand for the best technology made in the name of *cleaning up the air*; instead, clean air symbols were being manipulated in plain view on behalf of the *coal mining industry*. Nonetheless, Rogers's move prompted no serious discussion on the House floor—indeed, we do not know how many people were on the floor at the time the measure was introduced. After a discussion that occupies half a page in the *Congressional Record*, the amendment was adopted in an unrecorded vote.[9]

When a nearly identical measure was introduced in the Senate, however, it precipitated the only thoughtful discussion of the implications of the dirty coal–clean air coalition. The floor amendment, introduced by Senator Metzenbaum from Ohio on behalf of three other eastern coal-state senators,[10] was followed by a spirited debate in which eleven speakers revealed a clear grasp of some of the basic issues. The final interchange deserves to be rescued from page S9458 of the *Congressional Record*:

Mr. Muskie: [First,] the dominant thrust of this amendment is not its relationship to clean air, but its relationship to the economics of the areas it is designed to protect. For that reason I regret that it is being offered to a clean air bill. . . .

Second, I must say in all candor that the Constitutional Convention was the result of the fact that thirteen colonies were busily erecting trade barriers against each other. I do not think we want to get to another constitutional convention. I just see this as a first step, benign as its intentions are, as a first step in that kind of regionalization, around regional economic interests. It disturbs me a great deal. It is a propensity we all find in all of our regions and all of our states from time to time. For that reason, I will oppose it.

Third, insofar as its mechanisms are concerned, I think they could stand some refinement if, in fact, the Congress were finally to decide that the policy it represents is a good one.

Mr. Metzenbaum: I would like to respond and say to the distinguished Senator from Maine that I do appreciate the position he has taken and his fairness in doing so.

I want to point out that such prestigious groups who have concern with the whole question of clean air as the National Clean Air Coalition, the Environmental Policy Center, and the Sierra Club, have all come out formally in support of this amendment.

Those groups that are concerned with jobs, such as the United Mine Workers, who are directly concerned, the AFL-CIO, the United Steel Workers of America, the United Transportation Union, all of them have indicated their support of this amendment.

Last, but not least, ConRail has indicated its support.

Now all we need is the support of a majority of the Members of the U.S. Senate.

Note the artful way Metzenbaum parried the concerns of Muskie, the principal architect of environmental legislation drafted in the past two decades. Rather than concede an assault on the integrity of the Clean Air Act, he beckoned his comrades to join hands with the National Clean Air Coalition in support of the fears of the United Mine Workers.

While rhetoric is not everything in politics, the symbolic point takes on a new meaning in light of the roll-call vote that immediately followed Metzenbaum's plea. The amendment carried by a vote of 45 to 44—Metzenbaum collecting just enough eastern and southern Democratic votes to offset solid western support for Muskie's position.[11] Minutes after the amendment carried, the Senate took a step to control its potential for mischief. Senator Domenici of New Mexico gained unanimous support for a proviso that was accepted without debate or roll call. Implicitly acknowledging Section 125's status as special interest legislation, it stipulated that a governor, the president, or the president's designee "shall take into account the final cost to the consumer" before invoking the section.[12] This proviso made it clear that scrubbers should not be imposed if their cost was disproportionate to the protectionist benefits generated for the mines.*

When put to a test, then, the clean air–dirty coal coalition could gain only an equivocal and highly qualified victory. Any market guarantee for dirty coal could be obtained only through a highly visible procedure in which decision makers found a causal link between pollution requirements and "significant local or regional economic disruption" and recognized the heavy costs to consumers that unnecessary scrubbing might entail. Because there was reason to expect a substantial increase in demand for eastern coal over the next decade,[13] Section 125 would provide limited areas with temporary relief—and even this was not guaranteed. Confronting the question squarely, the Senate gave the dirty coal–clean air lobby more symbol than substance.

In contrast, its treatment of the same issue under Section 111 stood out in bold relief. The only unambiguous benefit to be gained by forcing all new plants to scrub, rather than use low-sulfur coal, was the same regional advantage that Senator Metzenbaum sought to procure. But here the future of forced scrubbing would not depend on a highly visible demonstration of economic dislocation in particular cases; it would be mandated on a nationwide basis at a cost of tens of billions of dol-

---

* In 1978, Congress further limited executive powers to issue regional coal use orders under § 125 and removed the power of a governor to initiate proceedings. *See* National Energy Conservation Policy Act of 1978, Pub. L. No. 96–619, 661, 92 Stat. 3206 (1978).

lars.[14] Yet, as the bill passed through the Senate, there was no easy way that anxieties about Section 111 could be expressed in statutory form. For the Senate bill, like its 1976 predecessor, proposed *no* change in Section 111, and the Senate report contained nothing resembling the House committee's low-visibility pronunciamento on behalf of dirty coal.[15] Moreover, the scrubbing lobby was content to evade explicit senatorial consideration and permit history to repeat itself—leaving it to the conference committee to reach a "compromise" in which the House's legislative history would once again expose the hidden meaning of Section 111.

MIDNIGHT LAWMAKING

The Senate's approval of the Clean Air Act on June 10 was only a preliminary to the main event—where conferees from both the House and Senate committees would hammer out a final compromise after two years of disagreement. As far as the main issues were concerned, compromise had become more difficult because of decisions made on the House floor to weaken PSD and to delay automobile compliance requirements. There were countless ancillary issues to be resolved, and precious little time for action. Massive layoffs might begin as early as September unless an acceptable auto solution were forthcoming, and much of urban America was threatened with a freeze on new industrial construction unless the 1970 act was revised.[16]

Nonetheless, the conferees reacted to these deadlines with a calculated nonchalance—although they were appointed in mid-June, they did not meet until July 18. This delay was even more remarkable in light of the House's announced intention to recess on August 5.[17] A Parkinsonian belief apparently prevailed: political dealing would expand to fill the time allotted and a final solution could only be reached a second before midnight. Indeed, the conference committee ended its hectic deliberations at 2:20 A.M. on August 3. Its final bill and report were typed during the day of August 3 and approved by the House and Senate the next day, so that these two bodies could adjourn on schedule.[18]

Once again, Section 111 occupied its familiar position in the congressional twilight. This time, however, a few glimmers of growing awareness can be detected. For Senator Domenici,[19] attending the conference of 1976 had had some educational value. Until that time, Domenici had been unaware of the House's invisible amendment. When its meaning was explained by the House staff, Domenici was immediately aroused and instructed his staff to prepare counteramendments and competing legislative history. The press of events made it impossible to formulate an adequate response in 1976, and the House's surprise tactics carried the day in the 1976 compromise bill. As a senator from New Mexico, Domenici came to the 1977 conference principally concerned with the future of PSD; but he would be better prepared this time to question Section 111. A similar gain in awareness was experienced by George Freeman, a principal spokesman for the utility's lobbying effort.[20] Once again, his main concern was to defend the House revisions of PSD, which had made the program more tolerable to western utilities. Over the summer, he came to recognize that the low-visibility amendment to Section 111 constituted a threat in its own right. During these critical weeks, however, Freeman could not afford to relax his efforts for PSD revision in order to mobilize an entirely new political campaign against universal scrubbing.

So when Section 111's turn came in the last week of the conference, the House amendment was not treated as an issue of the highest magnitude. When Domenici voiced his objections, the House proposal was not made the center of a day's discussion—though even this would hardly do justice to the issue in a room where a changing cast of harried congressmen and tired aides marked up proposals before a ring of watchful lobbyists of all persuasions. Instead of this first-class treatment, primary responsibility for Section 111 was delegated to a handful of congressional aides whose principals had expressed concern.

Because the House committee had taken an institutional stand, House staffers had the advantage of knowing the general position they were trying to advance. The situation on the Senate side was more complex. Although Domenici was a Republican in a Democratic Congress, his aide came to the bargaining session with substantial power, since the need to prevent a renewed filibuster made every vote important. But he could not press his views too far. The one thing the conference did not need was a major political confrontation on an issue that was considered of secondary importance. Also, Domenici did not have the solid support of his fellow senators. Senator Randolph of West Virginia, a coal state, was plainly in favor of the House amendment; the only question was how hard he would fight for it. Although Senator Muskie had shown himself unsympathetic to eastern coal interests in the vote on Section 125, Domenici's aide was uncertain if he would trade a concession on Section 111 for other issues he considered more important.

Within this setting of political uncertainty, a small group of staffers tried to work out a compromise version of Section 111. A series of legal uncertainties complicated their task further. Nobody could be sure what the House's committee report would mean in the cold light of litigation. Domenici's representative feared that the courts would give too much weight to the legislative history, and the House staff wondered if it would succeed in using that history to inflate minor changes in statutory language into a forced scrubbing requirement. Because Section 111 would no longer float through the conference unnoticed, the House staff was not averse to raising its hidden agenda to statutory prominence. Although both sides saw the merit of reworking the language of the House bill, an unambiguous statement of policy would precipitate the confrontation that all sought to avoid.

Senate staffers were successful in making one substantive point. A new subsection (h) makes it clear that Congress does not want the administrator to preempt the discharger's choice of the control technology that will best permit him to achieve emission limitations. Unless it is "not feasible" to permit polluters to choose, the subsection denies the administrator the authority to require a particular "design, equipment, work practice, or operational standard." The new definition of "not feasible" makes it plain that facilities like power plants cannot be sub-jected to design or equipment standards.[21]

While the Senate achieved this substantive prohibition, the House achieved a formal victory that leaned in the opposite direc-tion. Henceforth the statute would require the administrator to regulate power plants differently from other dischargers.[22] Al-though the administrator is simply to tell everybody else to reach specified emission limits, an acceptable power plant standard also requires:

> the achievement of a percentage reduction in the emissions from such category of sources from the emissions which would have re-sulted from the use of fuels which are not subject to treatment prior to combustion.[23]

Given this provision, the administrator must not only require a power plant to discharge no more than X pounds of sulfur ox-ide per MBTU, but also to reduce the sulfur in the coal by Y per-cent. By setting the percentage reduction requirement at a level only scrubbers can achieve, 85 percent for example, the admin-istrator could effectively force all coal burners to install scrub-bers. But the statute falls far short of mandating such a high per-centage. Indeed, it does not even require the administrator to establish the same percentage for all coal burners. Although each plant must be told "a percentage reduction" to achieve, an-other subsection, unchanged from 1970, expressly authorizes the administrator "to distinguish among classes, types, and sizes within categories of new sources for the purposes of establishing such [NSPS] standards."[24] Hence, the administrator has the au-thority to tell users of low-sulfur coal to reduce their sulfur con-tent by Y_1 percent and require high-sulfur burners to eliminate Y_2 percent.

Indeed, when viewed within the framework of the section as a whole, the new provision does not even bar the administrator from establishing a reduction of zero percent for low-sulfur coal burners. The new formal requirements are subject to the old substantive standard that requires the administrator to consider "the cost of achieving . . . emission reduction[s]" in defining the nature of the "best technological system."[25] This provision explicitly makes cost a consideration before low-sulfur coal users can be required to install scrubbers. Also, the statute's reference to the "best technological system" must be read in terms of the technical statutory definition, which includes "a technological process for production or operation by any source which is inherently low polluting or nonpolluting."[26] This expansive definition was introduced by the House in 1976, but the staff failed to modify it in the 1977 rush, even though it mocked their effort to prevent utilities from using "inherently low polluting" varieties of low-sulfur coal.

Rather than try to bring the increasingly complex statutory language under control, the draftsmen turned to their first love: making legislative history. Once again, the House staff gained a victory within the conference report that it failed to achieve on the surface of the statute: "The Senate concurs in the House provision with minor amendments. The agreement . . . preclude[s] use of untreated low sulfur coal alone as a means of compliance."[27] Yet midnight legislative history is a game any number can play. Aides for Senator Dominici were quick to add their own opinion in the next paragraph: "The conferees agreed that the Administrator may, in his discretion, set a range of pollutant reduction that reflects varying fuel characteristics."[28] Given this threat, the House staffers counterattacked with another tack-on: "Any departure from the uniform national percentage reduction requirement, however, must be accompanied by a finding that such a departure does not undermine the basic purposes of the House provision and other provisions of the act, such as maximizing the use of locally available fuels."[29]

These last remarks epitomize the abuses that follow last-minute staff-work. The only *statutory* recognition of eastern coal interests is found in Section 125, which does not seek to "maximize" local production but extends a highly qualified protection to areas suffering significant economic dislocation that can be demonstrably linked to pollution regulation. Rather than point to the coal lobby rhetoric contained in the old House committee report, an interpretation of the statutory language should refer to the aims of the statute declared in the act itself—which, apart from Section 125, shows no special solicitude to local coal producers. Just as the conference report fantasizes about the act's basic purposes, its invocation of a *"uniform* national percentage reduction requirement" finds no support in a statutory provision that merely requires "the achievement of *a* percentage reduction" to be determined after the administrator "tak[es] into consideration the cost of achieving such emission reduction" and allows the administrator "to distinguish among classes . . . within categories of new sources for the purpose of establishing such standards." This extraordinary bit of legislative history can be understood only as a desperate effort to offset the challenge raised by Senator Domenici's staff effort to "amend" the legislative history on its own behalf.

The incoherent quality of the legislative history was apparent to the amendment's supporters almost immediately. After spending all of August 3 compiling the conference report that had been completed at 2:20 that morning, the staffers thought it wise to spend the next day formulating a "Clarifying Statement" while both Houses were voting their approval of the statutory language. Among these clarifications is one that reflects the continuing effort to wrestle with the scrubbing confusion. This time it is said that

> while the conferees agree that the Administrator may set the percentage reduction requirement as a percentage range, the conferees expect the Administrator to be exceedingly cautious if he should elect to do so. Such range would be allowed only to reflect

> varying fuel characteristics and must be based on a carefully and
> completely documented finding . . . that [this] does not undermine
> the basic purpose[s] . . . of the House Report.[30]

But alas, the time had passed for putting such clarifications
where they belonged—in the text of the statute itself.

In short, the draftsmen brewed a mix of statute and legislative
history worthy of the occasion. Instead of integrating Section
111 into the basic structure of the act, their task was to avoid a
potential conference impasse by writing a document whose legal
meaning was hopelessly confused. The new Section 111 is easier
to understand as an exercise in small group dynamics than as a
serious effort to guide the bureaucratic management of a
multibillion-dollar problem.

CONGRESSIONAL BREAKDOWN?

Although we shall, in due course, speak in detail about the
proper judicial interpretation of Section 111, our concerns with
the congressional history go beyond this particular legal problem. For our story is rich with symptoms of acute institutional
breakdown that suggest general lessons about the effort to move
beyond the New Deal style of public administration. On the one
hand, our scrubbing story reveals a complete failure by the Congress to fulfill the functions it is uniquely equipped to handle in
our system of government. On the other hand, Congress was entirely inept in playing a role that could have readily been performed by an administrative agency.

Taking the first hand first, we understand Congress to be
uniquely equipped to discharge two governmental tasks: the *reconsideration of basic policy premises* and the *reflection of changes in
predominant political opinion*. But our long and complex story does
not reveal any effort to discharge either of these functions. At
no point did anyone use congressional reconsideration as a vehicle for ventilating the basic premises underlying Section 111.
Was it sensible to divorce new plants from old plants in the way

attempted by the 1970 act? Were there innovative ways of regulating new plants that would be more satisfactory than "technology forcing?" Even if the answers to such questions had amounted to an endorsement of the status quo, the ensuing policy discussion might have yielded regulatory rewards in years to come. Instead of making this effort, the House staff never looked beyond the issues generated by litigation to define problems worthy of congressional concern. Even within these narrow boundaries, there was no inclination to think through the basic policy issue raised by new coal burners—whether the administrator's 1.2 limit was too high or too low or just about right. Rather than refine the legislative definition of goals, it was necessary to avoid this question at all costs—lest it break apart the peculiar clean air–dirty coal coalition.

Nor did congressional reconsideration of NSPS provide an occasion for reflecting a change in dominant political opinion about environmental regulation. To the contrary, the incoherent congressional revision of Section 111 is a product of the special interests unleashed by the artless way the NSPS statute was written in 1970. The effort to insulate technology-forcing from New Deal ideals had succeeded only too well. Insulated from corrosive questions of means–end rationality, the symbols of environmental purity had been appropriated by—of all people—the partisans of dirty coal. Indeed the success of the scrubbing lobby depended entirely on its ability to avoid a well-focused debate on the basic issues raised by forced scrubbing. As we have seen, the only section that explicitly linked scrubbing to the protection of high-sulfur coal producers—Section 125—squeaked through the Senate by a roll-call vote of 45–44. Yet this highly qualified statute promised very little economic relief to eastern coal. Surely this vote did not augur well for a statute that would have explicitly required the expenditure of tens of billions of dollars on scrubbing so as to "maximize the use of locally available coals." Such cheap talk could survive in the committee reports only as long as it did not provoke a political confrontation that would force the issue into the center of congressional consciousness. Rather than reflecting prevailing political opinion, Section 111 could survive only by deflecting attention to other issues—PSD for the Congress, energy policy for the Carter administration.

Which leads us to the other hand. Rather than fulfill its distinctive functions, Congress operated like a peculiarly inept administrative agency, trying to resolve disputable issues of instrumental rationality without asking the most obvious questions raised by the scrubbing problem. What *was* so bad about sulfur oxide emissions anyway? Was forced scrubbing the cheapest way to solve the problem? Just how serious was the economic dislocation imposed by the shift to western coal? Were there better ways of cushioning the blow to eastern mines than spending billions of dollars on scrubbers? Rather than define goals with care and articulate cost-effective policies to implement them, the conference committee made its final decision among a crush of last-minute compromises on issues that seemed even more important. Yet we must move beyond these symptoms to more basic causes: how could the clean air–dirty coal coalition induce Congress to act—albeit incoherently—on a multi-billion-dollar issue without confronting the most obvious questions raised by the underlying problem?

At least part of the answer, we think, can be found in the political economy of congressional attention.[31] Each subcommittee recognizes that the capacity of the full Congress to process issues is such a scarce resource that only the most salient questions will receive serious attention from the great majority of representatives. Therefore each subcommittee is in a position to manipulate the legislative process to achieve aims that would not survive if they were given clearly focused legislative attention.[32]

The problem is heightened by the massive increase in congressional staff over the past twenty years.[33] Although these staffers do try to alert their congressmen to important issues, their presence also makes it possible for strategically placed interest groups to generate many more legislative initiatives. The demand for low-visibility legislation may be outstripping the fixed supply of congressional attention.[34]

Given this unhappy political economy, the case we have studied cannot be dismissed as aberrational. The fact that an obscure issue has not yet been subjected to hardheaded analysis can become a positive incentive for congressional action. As long as interest groups can provide superficially appealing symbols, they can hope that a single subcommittee's approval will not be challenged as their bill makes its way through Congress. This dynamic of agenda overload and symbolic subterfuge represents a challenge to all who wish to fashion a system of administrative law that represents a serious response to the admitted failures of the New Deal agency.

The second half of this book builds on our case study in an effort to suggest more promising ways to move beyond the New Deal. Rather than resolve disputed questions of instrumental rationality, Congress should be encouraged, with all the legal tools at its disposal, to clarify the controversial ends of environmental policy. By the same token, the EPA should not be allowed to suppress controversial questions of instrumental rationality by casually invoking the well-worn myth of expertise. Instead, it should be induced to resolve questions that require expertise in ways that will force bureaucrats to learn more about the complex areas they hope to regulate and will teach the rest of us about the costs and benefits of competing policies. At stake is the construction of an institutional framework that supports a sounder policymaking dialogue over time—one where Congress is obliged to reconsider values in light of emerging facts, and agencies are forced to clarify the costs and benefits of aims that captured the political imagination of earlier Congresses.

To lay the foundation for institutional reform, we shall examine our particular problem from four interrelated points of view: First, what does a singleminded concern with instrumental rationality reveal about coal scrubbing and the larger issue of sulfur oxides of which it is a part? Second, how did the EPA respond to the congressional initiative of 1977? Third, what questions should courts ask when called upon to review EPA decisions generated by congressional efforts to move beyond the New Deal? And finally, what direction should Congress take the next time it is forced by events to reconsider the Clean Air Act?

TO WHAT END?

A THOUGHT EXPERIMENT

In the first half of this book, we have shown how very different actors could come to view the scrubber as a plausible solution to their political problems. For eastern coal, the scrubber secured markets against western competition. For environmentalists, it promised to provide additional protection to pristine areas in the West. For the president, the scrubber assured political support for his high-priority energy program. For the conference committee, it was a nuisance that threatened to sabotage a long-delayed and vital agreement. All these interests converged to produce the legally incoherent message we have analyzed.

Although political cost-benefit analysis explains the outcome, there is another way of looking at the issue. What would universal scrubbing really accomplish in the world outside Washington, D.C.? Are the benefits worth the costs? What are the alternatives? Although a New Deal agency often answered these questions badly, at least it was expected to ask them. To pursue our interest in comparative institutional design, we shall look upon the scrubber as if it were a policy problem confronted by an ideal New Deal agency charged with the task of selecting the most sensible means to reach congressionally approved ends—protecting health and environmental quality by cleaning up the air. We will not try to pass judgment on scientific questions now anxiously debated among researchers. Instead, we will simply draw on the conventional wisdom now current in the EPA, to reproduce the way a knowledgeable EPA bureaucrat might analyze the problem if forced to argue in a means–end mode.

Fortunately, this exercise is not as much a flight into never-never land as it might seem at first. As we have seen, the Clean Air Act only goes part way in its rejection of New Deal models. As far as emissions discharged by old sources are concerned, the regulatory scheme is very much based on a New Deal attempt to design a pattern of cutbacks that will meet clean air targets fairly and efficiently. So we can find some solid ground for our thought experiment by considering the scrubber against the background provided by the EPA's efforts to deal with old sources. For these dischargers, the linchpin of the act is Section 109's requirement that the administrator establish an ambient standard for each serious pollutant at a level he deems sufficient "to protect the public health," allowing for "an adequate margin of safety." Once these standards are defined, each source is obliged to cut back its discharges so as to assure regional compliance with the ambient objective.[1] How, then, would the NSPS decision have seemed if the administrator had been obliged to take his own clean air targets into account?

DEFINING THE PROBLEM: SO_x

The answer, alas, is shockingly straightforward. The fact is that SO_2 concentrations in the United States have declined dramatically over the past twenty years.[2] As a result of this decline, most areas of the nation rarely experience violations of the clean air target the EPA has established in the name of public health.[3]

Imagine that, in the face of these "facts," a New Deal agency had tried to justify spending billions of dollars on scrubbing by pointing to the clear and present danger to health generated by sulfur dioxide alone. Such a move would have been reversed as "arbitrary or capricious" by any court of appeals in the land.[4] Yet this point is only the beginning of instrumental wisdom about scrubbing. For the "hard facts" collected by the EPA mislead as much as they instruct: before we can assess the scrubber we must ask ourselves what the EPA numbers measure.

One "hard fact" is that EPA's definition of the sulfur problem has failed to keep up with emerging scientific understanding. This is hardly remarkable, given the conditions under which the federal government first set a health-related objective for sulfur oxides in 1969. At that time the prevailing level of expertise reflected the pitifully small investment the nation had made in understanding its pollution problem. A national monitoring system was virtually nonexistent, with most data—of uncertain reliability—coming from a variety of metropolitan and state systems.[5] Even more important, the data typically measured concentrations of sulfur dioxide, although knowledgeable people recognized that SO_2, taken by itself, was an imprecise measure of the problem caused by a variety of sulfur oxides.[6]

Rather than the product of comprehensive study, the conventional emphasis on SO_2 was a product of the famous "killer fogs" of London and Donora.[7] In search of an explanation for these dramatic events of the 1940s and 1950s, epidemiologists made use of available meters to isolate probable suspects—the monitors rewarded this search by revealing relatively high concentrations of SO_2 *and* particulates in the air of suspect urban regions in the United States and Europe.[8] The 1969 sulfur oxide document recognized, however, that "because sulfur oxides tend to occur in the same kinds of polluted atmosphere as particulate matter, few epidemiologic studies have been able adequately to differentiate the effects of the two pollutants."[9] And as scientific activity increased in the 1970s, it typically generated no evidence of SO_2's harmful effects. By 1978, the National Academy of Sciences could report: "The notion that ordinary concentrations of sulfur dioxide alone are not likely to injure the lung is commonplace."[10]

Scientific inquiry then turned to more complex mechanisms in the search for a link between sulfur oxides and public health. The present focus is upon the production of sulfates: SO_4, not

SO_2; moreover, SO_4 is not considered by itself but in terms of the substances with which it combines to form tiny particles. For example, SO_2 may be transformed over time into a number of potentially harmful sulfate compounds, including sulfuric acid. Although larger particles are blocked in the respiratory passage before reaching the lung, micron-sized sulfate particles can elude these defenses and may be breathed deep into lung tissue.[11] The long-run harm, if any, caused by present concentrations of these tiny particles is far from clear, but the question is a central focus of present scientific concern.[12]

Unfortunately, EPA's regulatory program has simply failed to advance with scientific understanding. This failure is reflected in the data being collected. While a nationwide monitoring system now exists to test for compliance with the SO_2 standard, data on sulfates are far more fragmentary. It is even unclear if the glass filters EPA used until recently permit a reliable measurement of sulfate compounds.[13] And present filters are not analyzed for sulfates as frequently as they are for other pollutants —apparently because the chemical analyses required are relatively costly. Instead, the sulfate measurements we have are the result of a small number of independent research projects sponsored by or coordinated with the agency.[14] Much of this data may be too gross for an adequate assessment of health effects—for there is no reason to think that all sulfates are alike in their impact on lung tissue.[15] Nonetheless the data collected by EPA suggest the massive way the health problem must be redefined once the focus of concern shifts to include respirable sulfate particulates as well as SO_2. In contrast to the sharp reduction in SO_2 concentrations since the early 1960s, SO_4 concentrations generally have remained constant or have slightly increased.[16] Even more important, figure 1 dramatizes the uneven distribution of SO_4 in the United States.

The health threat, such as it may be, concerns a region east of the Mississippi and north of Tennessee and South Carolina. Average sulfate concentrations in *rural* areas in this region, which

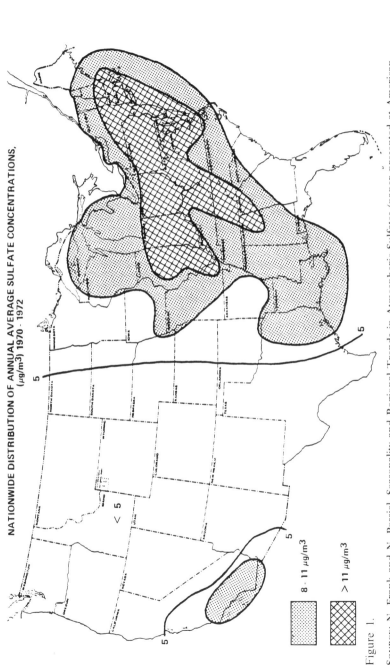

NATIONWIDE DISTRIBUTION OF ANNUAL AVERAGE SULFATE CONCENTRATIONS, (μg/m^3) 1970 - 1972

8 - 11 μg/m^3

> 11 μg/m^3

Figure 1.

SOURCE: N. Frank and N. Possiel, Seasonality and Regional Trends in Atmospheric Sulfates (paper presented at American Chemical Society, San Francisco, Cal., Aug. 30–Sept. 3, 1976) (figure 2). Reprinted by permission of the authors.

we shall call the Northeast, are often twice the typical level found in *urban* centers outside the Northeast. Within the Northeast, the typical city does not experience vastly higher sulfate concentrations than parts of the countryside. A similar picture is revealed if we turn to measurements taken during periodic episodes when, as a consequence of an air inversion, sulfate levels greatly increase. Once again, concentrations in the Northeast are higher during these episodes than in other regions.[17] If the health risk is generated by episodic exposures to 80 μg/m^3 as well as a steady exposure to 15 μg/m^3, the danger zone is not significantly altered.

This pattern comes as no surprise, given the gradual way SO_2, discharged from smokestacks, combines with other atmospheric ingredients to form sulfates. Instead of imposing the primary sulfate burden on the immediately surrounding area, each power plant makes an incremental contribution to a regional sulfate problem.[18] Moreover, the plant's impact on the regional problem is inversely related to its impact on the local sulfur oxide problem: the more sulfur oxide that drops out within miles of the plant, the less remains for gradual transformation into sulfates over longer distances.[19] Nonetheless, the EPA continues to define power plant compliance by measuring the SO_2 burden the plant imposes *locally*.[20] Given present concerns, this regulatory concern is downright perverse—creating a systematic bias against strategies, such as building shorter smokestacks, that will reduce a plant's contribution to the regional sulfate burden at the cost of higher readings on local sulfur dioxide meters.[21]

Such perverse enforcement schemes emphasize the opportunity lost when policymakers began to focus on technological means (the scrubber) rather than ultimate ends (public health). Imagine that, in the face of evolving scientific trends, the environmental movement had been obliged to pursue its objectives within a New Deal decisional structure. While there is no reason to think that this would have destroyed the environmentalists' love affair with the scrubber, at least it would have compelled a shift in argumentative strategy. To make the New Deal case for the scrubber, it would be essential to attack the EPA's continuing fixation with SO_2 and to point with alarm to SO_4 as the principal danger. For only then would a multibillion-dollar effort to reduce sulfur discharges below present levels seem plausible. Moreover, a successful redefinition of the principal pollution target would require a massive reorientation of the EPA's enforcement effort. At present, the EPA divides the country into 236 air quality control regions, each of which is responsible for forcing local polluters to meet local ambient standards. A rational approach to SO_4, in contrast, would require dividing the country into a smaller number of larger regions to take into account the realities of long-distance transport.

Any such change would encounter strong bureaucratic resistance. Yet with the policy debate focused on means rather than ends, nobody emphasized the need for overcoming this resistance. While there is evidence of congressional staff awareness of the emerging sulfate problem,[22] the clean air–dirty coal coalition had everything to lose and nothing to gain from a sober analysis of the bureaucratic implications of ecological realities. Rather than direct congressional attention to the need for a regional response to protect the East, the sulfate menace was used as a makeweight in the rhetorical case for forced scrubbing.*

PROTECTING HEALTH AND ENVIRONMENT IN THE EAST: THE PERILS OF SCRUBBING

Assume, next, that our hypothetical New Deal agency had man-

---

* While environmentalists emphasized forced scrubbing in their legislative activities, they did make an effort to use the courts to induce EPA to promulgate a sulfate target under Sections 108 and 109 of the Clean Air Act. Unsurprisingly, however, the judiciary proved unwilling to take responsibility for such a massive restructuring of agency effort, Sierra Club v. Train, No. 76–0656 (D.D.C. filed April 20, 1976; *dismissed with prejudice*, Jan. 19, 1978) (rejecting attempt to force promulgation of sulfate standard). Activist litigation is no substitute for thoughtful legislation.

aged to redirect its principal policy objective from SO_2 to SO_4 in line with prevailing scientific opinion. How would it go about appraising the threat posed by sulfates?

Although the impact of sulfates on humans remains uncertain,[23] there is little question of their harmful impact on some nonhuman species. Because of increased sulfur and nitrogen emissions, pH levels in rainwater have noticeably declined over the last several decades.[24] Long-distance transport from the Midwest and East creates the most acute effects in northern New York, New England, and eastern Canada. In freshwater lakes that serve as collectors of acid rain and storm water runoff, increased acidity has eliminated many fish species.[25] The effects of acid rain are most pronounced in geological regions where the characteristic rock types are highly resistant to weathering, with the result that acids are not effectively neutralized before reaching the lake ecosystems.[26] Many ecologists suspect that acid rain may significantly affect the structure and functioning of terrestrial ecosystems as well, possibly resulting in reduced timber and agricultural production. However, given the spatial and temporal variability in these systems, anything short of dramatic environmental consequences will be difficult to detect. There is cause for concern and room for further research.[27]

The first task, however, is to fashion an action program that responds sensibly to the dangers we perceive and allows us to define these risks with increasing precision over time. How does a scrubbing requirement for new plants fit into these larger goals?

Short-Term Action

A sensible short-term strategy for the Northeast must focus, first and foremost, on old plants, not new ones. For the next twenty years, the bulk of sulfur oxides in the Northeast will be produced by plants now in existence.[28] Moreover, SIPs in the midwestern states often permit old plants to emit four or five pounds of sulfur per MBTU. Since the 1971 NSPS held new plants to a 1.2 pound per MBTU ceiling, it is not obvious that short-term cutbacks can be purchased most cheaply by making new plants even cleaner. Unfortunately, the Clean Air Act of 1970 discouraged knowledgeable analysis of this trade-off between old and new plants; with old plants regulated through the New Dealish SIP process and new plants by an agency-forcing Congress, there was no forum in which the trade-off question might be considered as a part of the ordinary course of business.

Nonetheless, such an expert comparison would yield large rewards. While old plants can often achieve reductions by installing control equipment, important gains can also be obtained by washing sulfur pyrites from pulverized coal at the mine site. Although washing does not remove sulfur organically bonded to the coal, modern methods do remove 20 to 40 percent of the total sulfur content of coal currently mined in high-sulfur eastern regions.[29] At present, however, a leading EPA expert estimates that only 40 percent of eastern coal is washed before it is burned,[30] and EPA consultants estimate that universal washing could reduce eastern emissions by more than two million tons a year.[31]

The marginal cost of achieving this emission reduction seems far lower than anything attainable by forcing new plants to scrub. According to a leading EPA expert, the cost of washing high-sulfur coals ranges from two to nine cents per pound of SO_2 removed, compared to a cost range of seven to forty-five cents per pound for a 90 percent scrubbing system.[32] If the present sulfate threat justifies a serious short-term response, it is the less costly method—washing—that should be employed first.

All the more so because increased coal washing will yield sizable SO_2 reductions far sooner than scrubbing in new plants. Students of coal washing think it realistic to believe that at least a "1 to 2 million-ton" emission reduction is achievable by 1985.[33] In contrast, facilities regulated by the new NSPS will begin to come on line only in the late 1980s, and forced scrubbing in the

East will not achieve a remotely comparable impact upon the overall SO_2 load before the mid-1990s. Worse yet, by adding up to 15 percent to the cost of new construction,[34] forced scrubbing will give utilities an economic incentive to run their old plants longer than they might otherwise. Because old plants are often permitted to emit four or five pounds of SO_2 per MBTU, one old plant produces as much SO_2 as three or four new ones subject to a 1.2 ceiling. Even if a small fraction of old plants are induced to stay on line for an extended period, the overall impact could be quite serious. Indeed, as far as the industrial Midwest is concerned, the "old plant effect" swamps the extra reductions achieved by forcing all new plants to scrub—causing the Midwest to impose 170,000 *more* tons of SO_2 on the East in 1995 than it would have under the old 1.2 NSPS.[35] The old plant effect is not so powerful in other parts of the East. As a consequence, forced scrubbing does yield, by 1995, a net reduction east of the Mississippi of slightly less than one million tons when compared to the old NSPS.[36] Even this, however, is only a fraction of the "1 to 2 million" tons promised by washing in *1985*.

Coal washing, then, seems the short-term strategy of choice: it promises bigger gains sooner and more cheaply. How sad, then, that thanks to our effort to move beyond the New Deal, our policy has evolved in just the opposite direction. While policymakers rushed toward scrubbing in new plants during the late 1970s, they allowed the SIP process to remain focused on the attainment of local SO_2, rather than regional SO_4, air quality goals.[37] Old coal burners have been permitted to keep burning unwashed coal as long as their localities satisfied their SO_2 targets. An expert agency concerned with a cost-effective response to the emerging sulfate problem would have made the opposite choice.

Suppose, next, that our hypothetical agency concluded that the immediate sulfate problem was *so* serious that it made sense to cut the Northeast's present load by *more* than the "1 to 2 million" tons of SO_2 saved by increased coal washing.* Even on this policy assumption, however, forcing all new plants to install scrubbers would seem a silly way to achieve greater cutbacks. Rather than impose a high technology requirement on the utility industry, the goal should be to force utility executives to search out the cheapest way of achieving further cutbacks. If reductions can be achieved more cheaply by altering the operation of old plants or by buying low-sulfur coal, then it is these strategies that should be pursued. Moreover, the regulatory tools necessary to encourage utilities to search for the cheapest control technology are now very familiar in the pollution control literature.[38] A host of marketlike schemes have been proposed to force polluters to recognize the social costs of their emissions and to take efficient steps to cut them back. Under these systems, the burden of designing a cost-effective response is placed where it belongs—on the polluters themselves, not some government bureaucrats.

Even if an agency rejected such cost-effective regulatory systems and imposed a special ceiling on new plant emissions, forced scrubbing is plainly inferior to other methods. Instead of requiring the installation of scrubbers, it would be more sensible to reduce the emission ceiling required of new plants. Assume, for example, that the ceiling had been set at 1.2 pounds per MBTU. Then it is easy to imagine cases in which adding a requirement that polluters scrub 90 percent of the sulfur out of their coal yields absolutely *no* emission reduction. If cheap high-sulfur coal is readily available, cost-minimizing utilities may continue to discharge 1.2 MBTU and simply substitute higher sulfur coal for the more expensive low-sulfur varieties they might otherwise burn. But lowering the 1.2 ceiling guarantees lower

* EPA's consultants estimate that, in 1975, power plants emitted 16.0 million tons of SO_2 east of the Mississippi. Three million four hundred thousand tons of this total comes out of southeastern states whose contribution to the Northeast's problem is uncertain. *See* ICF, INC., *supra* note 28, at C–II–3a–b; *supra* note 36.

emissions from new plants, while inviting utility executives to define the cheapest way of meeting the new target. Given this contrast, no agency concerned with the cost-effective pursuit of clean air would have any trouble preferring a lower ceiling to forced scrubbing as a regulatory option.

Finally, forced scrubbing suffers a low-visibility disadvantage when compared to strategies that permit polluters to meet their obligations the natural way, by burning low-sulfur coal. A universal scrubbing requirement threatens to overwhelm the existing enforcement system; if this occurs, a symbolically satisfying gesture will disguise a very different operating reality. Once a scrubber is installed in a new plant, a utility will be free to buy high-sulfur coal on the expectation that it will be scrubbed down to the 1.2 requirement. If, however, the scrubber fails to operate, all the sulfur will go out the smokestack. The greater the efficiency demanded of the scrubber, the greater the pressure placed on the enforcement system. If the EPA requires 90 percent scrubbing and a utility achieves only 80 percent, the plant's emissions will be twice the legal limit.

The only way to assure compliance is through constant day-to-day enforcement activity. If high-level utility executives are quickly placed on notice as equipment failures arise, it is not too unrealistic to expect good faith efforts to obey the law.[39] But constant surveillance requires a reliable monitor in each smokestack, recording and reporting daily operations. These monitors are being developed, but their durability and reliability remain to be established.[40] Even more important, the EPA and state agencies must analyze the data promptly and dispatch trained inspectors to conduct spot checks when excess emissions are detected. Whenever a meter registers a potential violation, there are always two possibilities—either the scrubber or the meter is impaired. An official must then personally determine what is actually going on in the smokestack.

There is every reason to believe, however, that the EPA and the states do not intend to organize the necessary enforcement effort. Today, inspection visits are few and far between; usually the EPA has only unverified data submitted by the polluters themselves, and much of it is never entered into EPA computers for analysis.[41] As a result, the EPA has recently found that many systems thought to be in compliance are in fact violating present requirements.[42] Nonetheless, the bureaucracy intends to continue to rely on the polluters themselves to supply data on their scrubbing efficiency.[43] This means that the regulators will be unable to distinguish reliable from unreliable information; and without credible information, the threat of sanction is incredible.

Without constant enforcement pressure, it will be all too easy for the scrubbing operation to serve as the utility industry's Siberia—a place where employees unfit for money-making tasks are sent when it is inconvenient to fire them. The threat posed by incompetent or lackadaisical employees is especially serious since their tasks will not be restricted to mechanical maintenance operations. Scrubbers will constantly be demanding creative tending when they become clogged, corroded, or generally ornery. A conscientious, highly competent staff is an absolute requirement.[44] Given corporate incentives to place scrubbing on the back burner, it is especially important for the agency to create an administrative infrastructure equal to the challenge of enforcement.[45] Otherwise, the result will be a lot of junk in the smokestack and a lot more sulfur in the air.

In contrast, a strategy emphasizing the use of low-sulfur coals places much smaller burdens on the enforcement system. Meeting the 1.2 standard the natural way means that there is no need to monitor complex machinery on a daily basis. All that is required is a regular sampling of the sulfur content of coal going into the plant, rather than the smoke coming out of the smokestack. Although even this is beyond the present inspection efforts in many areas, it does not seem too much to hope that officials may, from time to time, be able to collect chunklets of coal for laboratory analysis.[46] When the realities of enforcement

are taken into account, low-sulfur strategies seem even more attractive than they appear in ordinary cost-benefit analysis.

Long-Term Responsibilities

But what about a cleaner tomorrow? Do we not owe something
to the easterners who will be breathing the air in the year 2000?

Certainly.[47] But once this is granted, the question of instrumental rationality remains to be asked: what steps should we
take today on behalf of the easterners of the next generation?

Begin, once again, with the obvious: we do not now know
enough about how sulfates of different kinds and quantities
harm us. There is still much to be learned about how much sulfate of each type exists, how sulfates are produced, and how they
are transformed during long-distance transport.[48] A generously
funded, long-term research operation is needed to fill the large
gaps in our knowledge. Yet the EPA is planning to spend only
fifteen million dollars a year on its Sulfur Research Strategy for
fiscal years 1980 to 1982, and an expenditure of only nine million dollars is projected for 1983—while energy consumers of
1995 will be asked to spend three or four billion dollars a year on
behalf of forced scrubbing.[49] This is a disgraceful failure to respond to the true interests of the next generation. A generously
funded, long-term research operation provides the only way we
shall ever clarify the health risks generated by alternative energy
sources.[50] Whatever measures we take in the short-term, surely
we owe this to our children. EPA's research budget should be
at least ten times its present fifteen million dollars. Even at
this price, it would cost no more than three medium-sized
scrubbers.[51]

Although research is a first priority, we must also recognize
that plants built today will last until the year 2020. It hardly follows, however, that immediate scrubbing is the best way to take
account of this fact. If the EPA maintained the 1.2 ceiling and
merely required present-day designers to plan for possible

scrubber retrofits, the marginal cost of adding scrubbers later would be relatively small.[52] With today's high interest rates, the loss caused by wasteful current investment is especially great. Moreover, twenty years from now technology may have moved far beyond the scrubber in its search for clean air.[53] At the very least, it will be possible to retrofit spanking new 2010 scrubbers into 1980 plants, rather than rely on creaky museum pieces. Why, then, act now?

Perhaps such a hasty, ignorant, and costly step might be morally required if a failure to scrub imposed risks on the next generation far greater than those that we ourselves must accept. The best available predictions, however, suggest that reckless disregard for posterity is not in question here. EPA models suggest that overall SO_2 loads in the East will be no higher in 2010 than they are today.[54] While the number of coal burners will increase, the replacement of 4-pound old plants by 1.2-pound new plants will completely offset the effects of increased coal use. Thus, even without scrubbing, we are not ruthlessly sacrificing the interest of the next generation to the convenience of this one. To the contrary, scrubbing is not only a costly way of providing the next generation with outdated machinery, but it will expose many northeasterners of the present generation to greater sulfur oxide concentrations than they would otherwise suffer.

Instead of relying on technological symbols, the task is to direct bureaucratic energy toward building ecologically sophisticated structures for sulfate control. The process of controlling sulfates should begin long before any smokestacks are built. The same discharge of SO_2 can generate different amounts of sulfate in different places depending on a host of geographic and meteorological conditions. A critical part of sulfate planning comes in judicious site selection—the more SO_2 that settles in nearby unpopulated areas, the less that remains for long-range transport in the form of sulfate. For example, if a power plant can be located in unpopulated hill country, this would go a long way toward reducing its impact on regional health. More broadly, prevailing meteorological patterns may make some regions of the country more desirable sites for power plants than others.[55]

Even after a site is selected, the design of the plant will influence its sulfate contribution. For instance, the higher the smokestack, the lower the nearby SO_2 readings will be, but the greater the plant's contribution to long-range SO_4. Although conventional wisdom now belittles the effect of low-level concentrations of SO_2, there is a point where higher concentrations will begin to affect asthmatics and other sensitive groups. Rather than deal exclusively with the scrubber in the smokestack, an intelligent long-range plan must try to define the extent to which the height of smokestacks can be reduced in the name of sulfate control without imposing unacceptable pollution levels near the plant.[56]

Finally, a bit of long-range planning may ease the chronic enforcement problem afflicting present-day control efforts. When a utility decides to build a new plant, it either builds a "captive" mine or contracts for a guaranteed supply of coal.[57] It is at this point—years before the first coal is burned—that enforcement agents may intervene. Steps should be taken to assure that the captive mine site will yield coal that will uniformly satisfy the Clean Air Act's requirements. These steps should ameliorate, if not solve, the problem posed by plants receiving shipments of coal that vary greatly in sulfur concentration.[58]

The next generation, then, will have just cause for resentment if we fail to resist the urge to scrub. Rather than leave our children some obsolescent machinery scattered in the smokestacks of the year 2000, we might hand them a worthier inheritance: a deeper understanding of the risks involved in coal burning; a subtle mechanism for siting and designing plants to do the least harm; and a mechanism for limiting the risks of predictable enforcement lapses. With such help from the present, the future will be in a position to take care of itself far better than we can protect ourselves today.

BEYOND THE EAST: THE VISIBILITY PROBLEM

As far as health and acid rain is concerned, the critical problem area is the Northeast—in the foreseeable future, sulfate levels west of the Mississippi will not approach typical eastern levels. And there is no reason to value the health of a westerner more than that of an easterner.

But as far as aesthetics are concerned, regional priorities must be reversed. Consider a bit of basic science dealing with the principal aesthetic problem—the danger that high-sulfur emissions will create a haze that impairs visibility.[59] Begin with earthly perfection—a condition in which the air is *completely* free of all particles—both man-made and natural. At this extreme, visibility is impaired only by the scattering of light caused by air molecules themselves—this "blue sky" or Rayleigh scattering establishes a maximum possible visible range of approximately 200 miles. As particles are introduced into the pure air, however, visibility declines at an *exponential* rate. Thus, where visibility is already impaired by the particle-producing activities of an urbanized and humid region like the Northeast, the marginal impact of a new plant on visibility is relatively small. If, for example, eastern visibility is assumed to be eight miles, then an increase in sulfate levels of 4 $\mu g/m^3$ will reduce it by less than half a mile. In contrast, if the air were perfectly particle-free, the same increase in sulfate concentration would reduce visibility by half—from about 200 miles to less than 100 miles.[60] Although no place is entirely particle-free, the Rocky Mountain states and the Southwest do enjoy large areas of extraordinary visibility. EPA presently estimates that median southwestern visibility is in a range between 65 to 80 miles, with visibility extending 10 percent of the time to 110 miles.[61] These facts not only require a change in geographic focus but a very different analysis of the power plant problem.

Power plants may pose two different threats to western visibility, requiring different methods of control. On the one hand, each plant contributes to the background level of particles detectable in a large region surrounding the plant—the background problem. On the other hand, each plant has a more concentrated impact on the narrow strip of territory downwind from its stack—the plume effect.[62]

Turning first to the plume, the visibility problem posed by sulfur oxides is not a matter of belching smokestacks. The only sulfur oxide emitted directly by power plants, SO_2, is a colorless gas. Visibility is impaired by the slow and long-distance transformation of this gas into sulfate particles. Thus, the plume has a perceptible impact on visibility only when it remains concentrated long enough for substantial sulfate formation to occur.[63]

For an ideal New Deal agency, the first step in mitigating the plume effect would be to shift enforcement efforts from average SO_2 levels measured over long periods to SO_2 levels prevailing on days when meteorological conditions permit the plume to retain its integrity for distances of fifty to a hundred miles.[64] Rather than spend billions of dollars to scrub tenths of pounds out of low-sulfur coal, it makes more sense to spend millions on a beefed up enforcement system that assures day-to-day compliance with visibility objectives. A rational policy for preserving visibility also would establish varying ceilings for plants of different sizes. The overall quantity of SO_2, not the amount of SO_2 per MBTU, determines the size of the plume. A 2,000 megawatt plant limited by a 0.6 pound per MBTU ceiling creates, ceteris paribus, the same plume effect as a 1,000 megawatt plant regulated by a 1.2 pound per MBTU ceiling.

Moving beyond present polluters, we need guidelines for building new plants. Smokestack height affects the concentration of sulfates and visibility hundreds of miles away. Moreover, tradeoffs between nearby SO_2 and distant SO_4 pose fewer dangers in the West than in the East. Because western power plants often are located in remote regions, they can use short stacks to enhance visibility without adverse health effects. Finally, better scientific models are urgently required to predict a future plant's plume effect. Such modeling, in conjunction with study of the meteorological conditions prevailing at potential sites, may reveal locations where plumes will diffuse with minimum impact on important clean air regions or populated areas.[65]

After a plume disperses, it adds to the background SO_4 level over a large region. In some parts of the West, particularly Arizona, diffuse emissions from a few large smelters have perceptible effects on visibility over large areas (even though background levels, by eastern standards, remain low). If we consider the Southwest and Rocky Mountain states as a whole, smelters produce several times the amount of SO_2 generated by the region's power plants.[66] Hence, the small incremental reductions obtained by full scrubbing would have little effect on background levels for the next generation. Conversely, even with less than full scrubbing, the EPA predicts that visibility in the region will improve by 1995 if smelters institute statutorily mandated emission reductions.[67]

Thirty years from now, power plants will play a larger role in the overall SO_4 picture in the Southwest and greater concern may prove justified.[68] Therefore, it may make sense to impose a retrofit design requirement. At present, however, the background problem cannot rationally justify billions of dollars spent on scrubbing, especially when the smelting industry is permitted—thanks to another 1977 amendment[69]—to delay compliance until the year 1988, and may succeed in deferring its day of reckoning yet further.

Finally, we must devote more resources to research on the facts concerning visibility. At present, we are not even confident of the basic equation which purports to explain how far the eye can see in a pollution-free world.[70] And it is plain that visibility is a complex psychological-aesthetic concept that must be measured with great sensitivity in a number of different locations.[71] Scrubbing will not substitute for this complex work.

Despite all this, forced scrubbing in the West does not look quite as silly as it does in the East. As far as the West is concerned, scrubbing is an exceedingly expensive way of achieving minor reductions in sulfates;* in the East, scrubbing is not only far more expensive, but may be positively counterproductive. Even on optimistic assumptions, forcing new plants to scrub will lengthen the life of dirty plants and may generate increases in emissions of sulfur oxides in the industrial Midwest for the next twenty years. And once a realistic view of enforcement is taken, forced scrubbing may make the sulfate problem worse, not better, for even larger portions of the nation's vulnerable northeastern quadrant. There must be better ways of moving beyond the New Deal.

* Some environmentalists apparently favor forced scrubbing of eastern high-sulfur coal in an effort to reduce the impact of strip-mining in the West. *See, e.g., Federal Coal Leasing Program: Hearing before the Subcomm. on Minerals, Materials, and Fuels of the Senate Comm. on Interior and Insular Affairs*, 93d Cong., 2d Sess. 275-79, 420-25 (1974) (statement of Katherine Fletcher, Environmental Defense Fund); *id*. at 484-500 (statement of representatives of several environmental groups, including Sierra Club).

It is not obvious, however, why the environmental damage caused by eastern mines should warrant less concern than damage in the West. In any event, forced scrubbing provides a singularly ineffective way of protecting the West. Even if scrubbers are installed, EPA predicts that western production will continue to boom, in part due to large increases in western consumption of coal. *See* ICF, INC., *supra* note 28, at A–lb to A–7b (showing difference in western coal production of 5 percent or less among alternative standards). Even if EPA retained the 1.2 standard, coal production in major regions west of the Mississippi is predicted to be less than 10 percent more than it would be under a full scrubbing standard. *Id*. at A–lb, C–II–9.

The best way to respond to the very real problems caused by strip-mining is through creative efforts at reclamation. Moreover, this need has been increasingly recognized both on the federal level, with the passage of the Surface Mining Control and Reclamation Act of 1977, 30 U.S.C. § 1201–1328 (Supp. II 1978), and the state level, with the enactment of increasingly high severance taxes and other devices to assure responsible reclamation practices. *See, e.g.*, 15 MONT. CODE ANN. § 35–103 (1979) (imposing 30 percent severance tax on coal produced in state by strip-mining); 82 MONT. CODE ANN. § 41–113 (1979) (establishing special fund for land reclamation).

6

EXPERTISE IN THE
SERVICE OF POLITICS

EPA AT THE CROSSROADS

We can now reconstruct the basic problem confronting the EPA when Congress delivered itself of its "clarifying" amendments on August 4, 1977. On the one hand, EPA policymakers could have looked on the incoherent last-minute statutory revision of Section 111 as an invitation to analyze the scrubber carefully by exploiting the substantial talent and insight present at the middle and lower ranges of the bureaucracy. On the other hand, policymakers could have looked behind the statutory language to the legislative history and sought to implement the will of the clean air–dirty coal coalition that had lobbied the change through Congress. Although the EPA ultimately adopted the second view of its mission, this was hardly the result of a single self-conscious decision. Instead, the rejection of New Deal ideals was the outcome of a complex process of bureaucratic struggle and political intervention.

AN AGENCY AT WAR WITH ITSELF

Almost immediately, the EPA found itself divided along lines that reflected Congress's problematic attempt to move beyond the New Deal. On one side, the Office of Air, Noise, and Radiation (Air Office), run by Assistant Administrator David Hawkins,[1] viewed Section 111 in political terms. From this view, the critical point was that the clean air–dirty coal lobby had succeeded in pushing *something* through Congress. Unless something spectacular was shown, it was wrong for an administrative agency to deny political activists the fruits of their congressional victory. Section 111, and especially its legislative history, had created a strong presumption for nationwide scrubbing, which was not to be offset by some technocratic mumbo jumbo.

On the other side, the Office of Planning and Management (Planning Office), headed by Assistant Administrator William Drayton, represented the technocratic view. This office is composed principally of economists and policy analysts professionally predisposed to considerations of cost effectiveness. They saw forced scrubbing as a pure waste of scarce economic resources. If scrubbing was the cheapest way of meeting the NSPS standard, polluters would scrub without the need of an EPA command. Only when cheaper ways existed would EPA's edict become significant. Moreover, the statutory language—in contrast with the legislative history—made it clear that the administrator must take "into account the cost of achieving . . . emissions reduction" before imposing *any* requirement under Section 111.[2] According to the Planning Office this meant that the agency should not follow through mindlessly on the congressional "decision" but should explore the scrubber's policy justification.

The conflict between the political and technocratic sides took shape early. As the bureaucratic home of the Office of Air Quality, Planning and Standards (OAQPS), Hawkins's Air Office had the inestimable advantage of setting the terms of intraagency debate by making the first proposal. During late fall of 1977, the Air Office circulated a recommendation for full scrubbing that required all coal burning plants not only to meet the old 1.2 limit but also to use scrubbers to remove 90 percent of the SO_2 released by the coal the plants burned.[3] In response, Drayton's Planning Office relied on a computer model to determine the costs of forced scrubbing.

Unfortunately, the Planning Office's modeling team defined

its mission in an extraordinarily narrow way.[4] First, the model-ers made no effort to use their multimillion-dollar exercise as a vehicle for refining the agency's understanding of the environ-mental benefits to be gained from scrubbing. All NSPS options were analyzed exclusively in terms of SO_2 even though the staff knew of the links between SO_4 and more serious environmental hazards.[5] Rather than use available bureaucratic expertise to redefine the problem explicitly, the modelers took only the most halting steps toward allowing the emerging scientific con-sensus to influence their construction. In addition to predicting the total tonnage of SO_2 produced on a nationwide basis under alternative programs, they divided the country into six regions—shown in figure 2—and predicted tonnage for each. Remarkably, these regions were defined without regard to the most basic meteorological facts. Thus, the critically important Ohio River Valley was split into two regions—following state lines rather than making any effort to highlight the region's ag-gregate impact on the Northeast's sulfate problem.[6] The seem-ingly innocent decision to provide a single number reporting nationwide SO_2 emissions was particularly insidious—since it failed to separate the very different interests at stake in the dif-ferent regions. Instead of examining impacts on health, acid rain, and visibility, the model spoke in terms of millions of tons of SO_2.*

* The modelers' crude breakdown of the country into six regions constrains the policy discussion that follows. As explained in chapter 5, policymakers should have been most concerned with the risk to health and ecology in the northeastern quadrant of the country identified by the map on page 63. Since the modelers defined their regions so poorly, however, we have been obliged to speak as if the emissions discharged anywhere in the three model-regions east of the Mississippi posed the same risk to the Northeast. This is a gross oversimplification; the contribution of the model's East South Central region to the Northeast's problem is especially unclear. Nonetheless, it seemed better to err on the side of over-, rather than under-, inclusion. Hence when we speak of the "East," we mean to include all of the country east of the Mississippi.
So far as visibility is concerned, the model's Mountain region seems most sa-lient. *See* pp. 75–78, *supra*. Unfortunately, many of the computer runs did not

Second, the modelers examined a very limited range of policy alternatives. They ignored the possibility of more sophisticated siting or design strategies. Indeed, they failed to explore *any* option that would allow *any* new plant to avoid the immediate construction of scrubbers. No consideration was given, for example, to the possibility of requiring 1990 plants to burn low-sulfur coal in the short term and install scrubbers in 2010, even though it was recognized that immediate scrubbing would induce utilities to keep their dirty old plants on line longer.[7] Instead, the modelers contented themselves with modest variations on Hawkins's full scrubbing theme. They focused exclusively on the possibility of allowing some scrubbers to scrub away something less than 90 percent of the sulfur in a plant's coal—thereby saving the operation and maintenance cost involved in intensive 90 percent use.[8] Finally, their model took an unrealistic view of enforcement. By assuming perfect compliance, it failed to alert policymakers to the danger that forced scrubbing might seriously worsen the Northeast's sulfate problem.

These analytic failures* are all the more remarkable in that

---

disaggregate Rocky Mountain loads from Pacific loads. When we speak of the "West," then, our numbers will include both Rocky Mountain and Pacific loadings. This is less serious than it would seem—since Pacific loadings are only about 25 percent of the total. *See* ICF iv, *supra* note 4, at C–II–3C; C–VIII–3C.

* The modelers' praiseworthy effort to take some important variables into account should not be entirely ignored. An effort was made to consider the sensitivity of model conclusions to changes in: the growth in demand for power, the amount of nuclear capacity constructed, the prices of oil and coal, scrubber costs, and coal transportation costs. When comparing alternative standards, the model's predictions are relatively sensitive to changes in some of these assumptions—especially the costs of oil and scrubbers. *See* ICF iii, *supra* note 4, at 29–31. Environmental groups strongly criticized assumptions used in the model as unjustifiably biased against full scrubbing. *See* NRDC & EDF, Comments, *supra* note 7, at V–1 to V–24. *See also* ICF iii, *supra* note 4, at 23, 36 (analysis of alternative standard using NRDC's preferred assumptions which reduced emissions and increased costs in 1990 relative to standards modeled using different assumptions). Even the computer runs favored by environmentalists, however, did not conceal the fact that forced scrubbing would cost billions more than a cost-minimizing strategy.

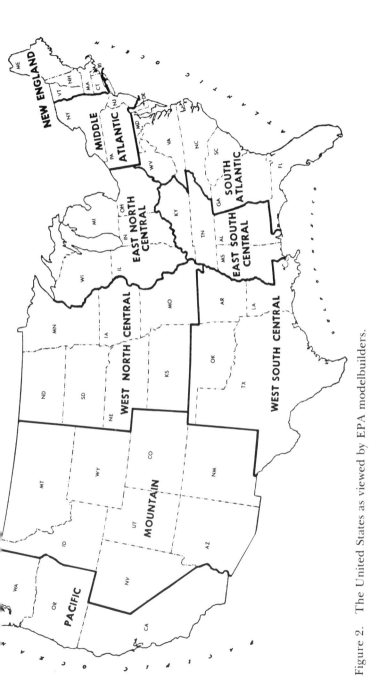

Figure 2.   The United States as viewed by EPA modelbuilders.

SOURCE: Map provided by the EPA.

they were well understood by most of the staff who worked with the model. They are not explicable in terms of ignorance but in terms of bureaucratic stimulus-response. Not only had the Air Office seized control of the agenda by making the first proposal, but the internal structure of the agency deflected the Planning Office from a more rewarding exploration of policy alternatives. The people at EPA best qualified to analyze environmental benefits were the agency's technical staff in North Carolina. Yet much of it was under the bureaucratic control of Hawkins, and it would be naive to imagine that he would appreciate efforts to arm his bureaucratic opponents. So the Planning Office looked outside the agency for much of its analytic work. It persuaded the EPA to commission a Washington, D.C. consulting firm known as ICF, Inc., to play a central role in the basic modeling effort. Although this firm was well known and highly regarded for its models of coal supply and energy production, its computer model did not even attempt to translate SO_2 emissions into SO_4 impact. Rather than encourage work on this vital environmental question, the model's detailed forecasts of national oil consumption and coal supply invited the vigorous participation of staffers from the Department of Energy, who saw the scrubber as a serious obstacle to energy independence.[9] The institutional stage had been set for a computer analysis rich in its discussion of costs but impoverished in its consideration of the benefits that putatively justified all the activity. As a result of the Air Office's political orientation, EPA would invest millions of dollars on an analytic enterprise that did not attempt a sophisticated advance in the environmental dimensions of the policy problem.

Despite its flaws, the model provided the Planning Office with a powerful weapon. Even when so many critical issues had been defined out of existence, the dubious virtues of full scrubbing could not be entirely concealed. The agency's final computer runs predicted that full scrubbing would only yield modest nationwide reductions in SO_2 as late as 1995.[10] In that year, full scrubbing would yield 20.6 million tons of SO_2—only a 13 percent drop from the 23.7 million tons that would be tolerated under the 1971 NSPS. This small reduction emphasized the relatively small role new plants would play in the total emissions picture as well as the possibility that eastern plants might switch to higher sulfur coal if forced to scrub. Regional breakdowns made the picture even less attractive. They showed that SO_2 loadings would slightly increase in the Midwest, thereby illustrating the new incentives that scrubbing would give utilities to keep their old dirty plants on line longer.[11] Since policymakers were aware of the acid rain and health problems generated by sulfates in the Northeast, this regional news—however unsophisticated—was not heartening. In contrast to this gray picture on the benefits side, the model revealed that full scrubbing would be one of the costliest ventures ever imposed in the name of a cleaner environment. In order to move beyond the emission reductions attained by the old 1.2 standard, full scrubbing would cost an extra 4.1 billion dollars a year by 1995 and much more in later years as more new plants were built.[12]

While the computers clacked on, the agency began to confront yet another congressional constraint on its discretion. As part of its last-minute activity, the conference committee had imposed a one-year deadline for the EPA to implement the revised version of Section 111.[13] When it became clear in July 1978 that regulations would not be proposed, let alone promulgated, by that deadline, the Sierra Club successfully obtained a court order requiring an EPA decision by June 1979.[14] This timetable obliged the agency to publish a tentative proposal promptly in the *Federal Register*, so that interested parties would have the chance to submit their comments before a final EPA decision.

The scrubbing issue began to loom large on Administrator Costle's personal agenda during the summer of 1978. Intraagency conflict had the great merit of forcing the administrator to give more time to the issue than had any previous policymaker of like stature—either in Congress or the executive branch. Yet the dispute reached Costle through the distorting institutional prism bequeathed by the congressional effort to move beyond the New Deal. On the one hand, the Air Office kept its eyes firmly fixed on the legislative history and so believed a sophisticated analysis of environmental consequences unnecessary for its extreme position on behalf of full scrubbing.[15] On the other hand, the Planning Office, together with its consulting firm, expatiated on the costs with only the meagerest institutional capacity to speak intelligently on the question of whether the costs were worth the benefits. Rather than a coordinated institutional effort to define short- and long-term strategies to cope with the sulfate problem, Costle was presented with dollar figures from one faction and clean air rhetoric from the other.

EXECUTIVE POLITICS

In deciding on the proposal that the EPA would announce in the *Federal Register*, Costle was hardly operating in an insulated bureaucratic environment. From the time the Air Office circulated its first internal memo on behalf of full scrubbing, outsiders were informed of agency policy through formal, interagency contacts[16] as well as by agency leaks and the *Environmental Reporter*, a weekly publication of the Bureau of National Affairs. Thus, outsiders could intervene informally even before the EPA proposal was published on September 19, 1978.

The critical mover at this stage was the Department of Energy (DOE). Now that the scrubbing issue had been isolated from the swirl of comprehensive energy planning, the DOE bureaucracy began to assert its doubts about forced scrubbing. The participation of DOE staff in work on EPA's Planning Office model not only solidified staff doubts but produced reams of computer printout by which DOE bureaucrats could educate their titular superiors. Within months, Secretary Schlesinger had become convinced that his initial acceptance of forced scrubbing had been a serious mistake—making the shift from oil to coal more costly without compelling environmental benefit. And as Costle came to his time of decision in the summer of 1978, DOE made known its opposition to the Air Office proposal in increasingly aggressive ways. A letter from John O'Leary, Deputy Secretary of DOE, gained general circulation by July. It formally announced DOE's opposition to full scrubbing and endorsed a partial scrubbing alternative that was being developed with the aid of the Planning Office model.[17]

By August, DOE opposition forced Costle to the White House to hammer out an agreement on the proposal to be published in the *Federal Register*. In a memorandum presented at the meeting, Costle asserted that the new agency-forcing statute contained a presumption favoring full scrubbing but admitted that further analysis could overcome the presumption.[18] Rather than choosing sides in an authoritative way, Costle permitted the agency's formal notice to reflect the bureaucratic impasse. The proposal presented in the September notice was the Air Office requirement that all new plants reduce sulfur content by 90 percent as well as comply with the 1.2 ceiling—which was retained unchanged.[19] But the proposal also announced that "[t]he Agency believes that it would be inappropriate to make a decision on the choice between the full and partial control alternatives without additional analyses of the modeling results."[20] The ambivalence apparent on the face of the agency proposal reinforced Costle's earlier public declaration that his mind remained open.[21] Scrubbing would finally receive the high-visibility attention it deserved.

TECHNOCRATIC ASCENDANCY

Their work now acknowledged by the administrator, the computer modelers proceeded in earnest and produced an alternative to the Air Office's full scrubbing proposal. Since the Planning Office had already greatly narrowed its inquiry, it could not hope to come up with an environmentally sophisticated response to the emerging sulfate problem. While Planning proliferated program options over the next few months, *all* the options that survived to the final set of computer runs projected nationwide SO_2 emission loadings in 1995 that fell within 3 percent of the 20.6 million tons of SO_2 envisioned under the Air Office's initial full scrubbing proposal.[22] These small differences, however, did not prompt the Planning Office staff to force an intraagency confrontation on the key issue—whether an increase in the stringency of NSPS was a cost-effective way of responding to the threat of acid rain in the Northeast and haze in the West. Instead, staff energy was consumed in the elaboration of one well-known theme that could be played in countless variations.

The Rise of Partial Scrubbing

It is the siren song of Pareto-efficiency. It is child's play to show that any emission reduction achieved by 90 percent scrubbing can be achieved more cheaply by more intelligent regulation. As we have seen, if eastern coal burners were forced to scrub at 90 percent, they can sometimes respond by buying cheap high-sulfur coal, thereby offsetting the scrubber's emission reduction by increasing the total sulfur content of the fuel consumed. This simple point led the Planning Office to a sliding-scale approach to emission reduction. For example, if the utility promised to discharge only 0.8 rather than 1.2 pounds of sulfur oxide per MBTU, then the EPA might agree to forego its 90 percent scrubbing requirement and permit some lower level of percentage removal, say 33 percent. Under this sliding-scale option, the environment would gain by means of the reduction from 1.2 to 0.8, and the utilities would gain because scrubbing at 33 percent is cheaper than scrubbing at 90 percent.[23] The only folks who could possibly lose were the sellers of the very dirtiest coals. Moreover, since the Planning Office model was relatively sophisticated in its treatment of coal production, it could predict confidently that partial scrubbing·options would do no serious economic harm to high-sulfur coal. Indeed, major increases in overall eastern coal production were anticipated.[24]

There was, however, one interest that could not be paretianized by the use of the sliding scale. In the West, a 90 percent scrubber did imply a 90 percent reduction in new plant emissions since only low-sulfur coal was readily available. Some gains from 90 percent scrubbing would be offset by increased use of old plants,[25] but the model still predicted a modest reduction in western SO_2 loadings by 1995.[26] Modelers seeking Pareto superiority therefore made a regulatory distinction between East and West: easterners might be allowed the benefits of the sliding scale, while westerners would receive the full rigor of full scrubbing. Here, finally, was something that made nobody worse off than they would be under the Air Office's full scrubbing proposal and that had the advantage of providing both branches of the EPA with a face-saving compromise. Although most of these "eastern sliding scale—western full scrubbing" proposals were Pareto superior to the full scrubbing proposed by the Air Office, they only served to emphasize the bizarre quality of the entire exercise. After all, it was the *East* that suffered more severely from acid rain and the possibility of health impairments. Nonetheless, the East–West split would permit many eastern plants to keep discharging at 1.2 when almost all westerners were discharging at 0.2![27]

Although this bizarre solution did not force a reconsideration of basic premises, it did lead the technocrats to favor a competing proposal that would tolerate some increases in SO_2 in the West while reducing eastern loads more significantly. Moreover, this solution had the virtue of administrative simplicity and served to emphasize the economists' main insight. Why not simply reduce the old 1.2 emission ceiling to some lower number, say 0.55, that would force everyone to scrub somewhat since no coal could pass the lowered ceiling without advanced technology?[28] Despite this radical reduction in the NSPS ceiling, costs would be reduced as long as polluters were allowed to decide for themselves how to mix low-sulfur coal and higher scrubbing percentages to reach the 0.55 ceiling. Indeed, the computer suggested that these cost savings were *so* great that by 1995 a simple 0.55 limit would be cheaper than the Air Office proposal by 800 million dollars a year (4.1 billion dollars versus 3.3 billion). At the same time, nationwide SO_2 emissions would go from 20.5 under full scrubbing to 20.3 under the lowered ceiling.[29] These numbers looked even more attractive when broken into regional components. Compared to full scrubbing, the low ceiling option would reduce eastern loadings by 740,000 tons in 1995, from 15.6 million to 14.9 million. In contrast, by 1995 power plant emissions in the West would rise only by 280,000 tons, from 0.9 million to 1.2 million.[30] At least this overall pattern made sense: when compared to full scrubbing, lowering the emission ceiling to 0.55 yielded savings of almost a billion dollars a year, lower SO_2 emissions nationwide, and emission reductions in the areas east of the Mississippi where the health threat was most serious.*

By January 1979, opinion within the Planning Office was converging strongly upon this simple ceiling reduction proposal. Rational analysis had paid off: no serious person could think that the Air Office's proposal for full scrubbing was superior to the expedient of lowering the ceiling to 0.55. And it was at just this moment that the technocrats in the agency gained important reinforcement from on high, as the White House staff added its weight to their counterattack.

---

* For expository convenience, the text emphasizes emissions in the two parts of the country that raise the most significant policy problems—health and ecology in the East and visibility in the West. Between the West and the Mississippi lies the model's Central Western regions. See the map on page 83, *supra*. In these regions, a low ceiling would increase 1995 emissions by 320,000 tons (from 3.9 to 4.2 million tons) when compared to the full scrubbing alternative. ICF III, *supra* note 4, at D–III–1, –2. It is difficult, however, to link this increase to any interest as significant as those at stake in the other regions, and so we will not be emphasizing the Central Western region in our discussion.

The View from the Center

In the early days of the Carter administration, center stage was occupied by Schlesinger's heady energy program; by 1978, however, the scrubber was worthy of attention in its own right. As White House staffers became familiar with their offices, they tried—as have so many of their predecessors—to gain control over the sprawling bureaucracy around them. The organizational initiative of greatest interest here is the White House effort to implement a 1978 executive order requiring the agencies to conduct an "economic impact analysis" for every "significant" regulation. The core of this effort is the so-called Regulatory Analysis Review Group (RARG)—a bureaucratic box in which executive office economists confront ambassadors from cabinet-level agencies.[31] The RARG job is to review the analysis generated by the front-line agency in defense of its proposed initiative and to publish a report of its findings for inclusion in the agency's rule-making record.[32] Although this makes the RARG potentially significant in the practice of judicial review, our concern here is with its direct impact on policymaking. White House staffers on the RARG could credibly threaten to appeal to the president if the reviewing agency proved entirely unresponsive to their suggestions. While the agency ambassadors could choose to stand their ground and have their boss make the case to Carter himself, this was obviously a last resort. As in the conference committee situation on Capitol Hill, the job of the RARG staffers was to reach an agreement that their superiors could live with without undue discomfort.

The organizational heart of the RARG is a group of economists formally attached to the Council of Economic Advisors (CEA) and the Council on Wage and Price Stability (COWPS). This elite corps tries to make up for its small numbers and multiple responsibilities by a combination of energy and ability. Nevertheless, it was clear that, especially in the beginning, the RARG would review a very small number of regulatory proposals.

NSPS was only one of nine proposals given the RARG treatment in 1978. While its multibillion-dollar annual price tag made it a naturally attractive candidate to the cost-benefit analysts on CEA–COWPS, extra-White House realities helped to assure NSPS a place on the RARG agenda. By the summer of 1978, the scrubber had forced its way into the Executive Office Building from at least three directions. First, DOE's growing opposition to full scrubbing required, as we have seen, White House mediation in August 1978 before any EPA proposal could be formally issued.[33] Second, at about the same time, a leading utility representative was meeting with Stuart Eizenstat, presidential assistant for domestic affairs, to express his strong opposition to the Air Office proposal. Third, the Business Roundtable had targeted full scrubbing as one of the most obnoxious regulatory initiatives of 1978.[34]

These elements combined to make NSPS an ideal candidate for RARG review. Not only did the scrubber's high costs attract technocratic attention, but DOE opposition suggested the need for central coordination, and the opposition of the business community signaled the presence of political interests not readily expressed within EPA. For all this, White House staff "intervention" initially involved dispatching Robert Litan, a junior lawyer-economist at CEA, to spend most of three months on the scrubbing issue. By late December, Litan was sufficiently familiar with the Planning Office modeling effort that he could write a RARG draft along with other staffers under the supervision of William Nordhaus, a senior member of the CEA.

The RARG report[35] produced in this ad hoc way suggests the value of review by a technocratically competent staff working outside the control of the operating agencies.[36] Although hurriedly written to meet the January 15 deadline for submission to the EPA's public docket, it brought to the surface some of the larger questions that had been suppressed by the bureaucratic struggle within EPA. Most important, the report focused on the peculiar fact that EPA had failed to illuminate the *environmental* dimensions of the problem in a sophisticated way. RARG challenged EPA's concentration on SO_2 rather than SO_4.[37] Moreover, it emphasized the foolishness of examining total pounds of SO_2 emitted without taking account of the number of people actually exposed to emissions. Since, as we have seen, the Planning Office model suggested that full scrubbing increased emissions in the heavily populated Midwest, it was easy to construct a primitive "exposure index" that emphasized the trade-off between eastern health and western visibility implied by full scrubbing.[38]

Not that the RARG document made a systematic effort to measure scrubbing against the various goals—human health, ecological integrity, aesthetic visibility—relevant in a comprehensive analysis. Nor was there a challenge to the idea that *all* new power plants must *immediately* install scrubbers—although there were passing allusions to the importance of old plants and the possibility that SO_2 reductions could be purchased far more cheaply by controlling them, as well as other industrial sources, more rigorously. Instead of emphasizing these broader points, however, the RARG focused on the battle already raging within EPA—coming out clearly against full scrubbing and in favor of Planning's proposal to lower the emission ceiling from 1.2 and to permit utilities to mix scrubbing and low-sulfur coal in the way that would most cheaply meet the lowered ceiling.

All in all, then, RARG did a remarkably good job with the resources at its disposal. It discharged the most fundamental reviewing task—challenging the agency's unduly narrow definition of its problem and pointing out concrete ways the problem might be usefully redefined. And it clearly placed the burden of justification on the side that seemed to have the worst of the technocratic argument within the framework provided by the agency's existing definition of the problem. Unlike judicial review, RARG review came at a time before bureaucratic positions had hardened beyond alteration. The RARG report was filed on the last day the EPA's September proposal was open for public comment. The time had come for Administrator Costle to make up his mind.

AGENCY DECISION

The RARG report *"urge[d] EPA to analyze exposure and health effects before reaching its final decision on the form of the NSPS."*[39] Yet it was already plain that Hawkins's Air Office would refuse this invitation; from the very beginning of White House staff intervention, the Air Office had rejected such requests as "counterproductive."[40] While the Air Office refused the call to make use of its latent expertise, the RARG's more narrowly focused rejection of full scrubbing on behalf of a lowered ceiling had more impact. Rather than vainly trying to transform bureaucratic realities, this message reinforced powerful opponents of full scrubbing. Not only did it enhance the internal credibility of EPA's Planning Office, but it also made plain that other bureaucracies—notably, DOE—would have significant support in the White House if the EPA entirely ignored their opposition.

The High Tide of Technocratic Rationality

As the public comment period closed on January 15, 1979, the agency's perception of NSPS had matured significantly in the year since Hawkins's Air Office first proposed its full scrubbing–high ceiling option. The technocrats had plainly made a strong case for a partial scrubbing–low ceiling strategy, and if this initiative were to be deflected, something new would have to be added to the bureaucratic equation.

This "special something" was precisely what Hawkins's Air Office tried to provide. As far as a committed environmentalist was concerned, there was nothing wrong with the idea of lowering the ceiling from 1.2 pounds to 0.55 pounds. But it would be even better to add a full scrubbing requirement on top of this, thereby reducing western emissions even further to 0.2 pounds of SO_2 per MBTU. Thus, while the Planning Office proposal might be better than the original full scrubbing–high ceiling proposal, there still remained room for a new Air Office initiative on behalf of a *low* ceiling with full scrubbing.

There was only one problem. While it was plain that the reduction of western emissions from 0.55 to 0.2 would lower the western tonnage figure appearing on the computer printout, it was not obvious that this numerical triumph implied a real environmental gain in the world beyond the computer.[41] Thus, in a wonderful turn, the Air Office began conducting a benefit analysis early in 1979, after refusing express White House requests in the previous summer. Without circulating its work to the Planning Office in the normal way,[42] the Air Office tried to show that the reduction from 0.55 to 0.2 in the West would make a real difference in environmental quality. While Planning Office computer projections concentrated on the year 1995, the Air Office focused on 2010 or later. By looking far into the future, the Air Office computer runs magnified the significance of the difference between 0.2 and 0.55 in the West. Its model runs predicted that, by 2010, western power plants would generate 4.4 million tons of SO_2 under a 0.55 limit, while only 3 million tons would result under a 90 percent scrubbing standard.[43] Hence, according to the Air Office, the need for full scrubbing had been reestablished on firm technocratic ground.

The unreality of these projections, however, was easily unmasked when they came to the Planning Office's attention in February 1979.[44] Because NSPS would be reconsidered before 2010, it was wrong to ignore the possibility of tightening up more in the future even if the 1980 NSPS were "only" set at 0.55. Moreover, in its zeal to establish the benefits of full scrubbing in 2010, the Air Office had failed to estimate the costs imposed on new plants built between 1995 and 2010. These added costs were later placed at 6 billion dollars a year by 2010.[45] Finally, the Air Office failed to note the contribution of smelting to western sulfur loadings. Because the Planning Office expected smelters to reduce their discharges substantially over the next half century, it argued that overall SO_2 loads would remain constant, or even decline,[46] under the 0.55 limit. These counterarguments effectively undermined the Air Office's belated effort to defend itself with the same flashy weapons brandished by its opponents.

A related clean air effort, however, did manage a brief moment in the sun. During February, the EPA circulated a partial scrubbing–low ceiling proposal to other executive departments for comment.[47] In response, the Department of Interior (DOI) expressed its concern about future visibility in the western national parks in its bureaucratic bailiwick.[48] Because DOI's mission did not oblige it to take into account any of the scrubber's costs, there was nothing to prevent it from pronouncing itself in favor of full scrubbing. Not content with a paper statement, Secretary Andrus met with Administrator Costle to express his position personally. Shortly thereafter, DOI made a formal presentation to somewhat skeptical analysts from the EPA's Air Office,[49] dramatizing Park Service concern with a slide show that depicted a hazy Grand Canyon and belching smokestacks (on separate slides). Unfortunately, DOI was unable to establish any causal link between the two halves of its slide show. As we have seen, the visible portion of a smokestack's plume is not the product of SO_2;[50] moreover, given present emissions, it is Arizona's smelters, not coal burners, that are more likely to be responsible for a hazy Canyon—if the haze is to be attributed principally to man-made sources.[51] While Air Office staffers questioned the slide show at its first screening on March 1, a group was detailed to see whether something could be made of the visibility issue—especially given the Air Office's inflated SO_2 projections for 2010. Within a couple of weeks, the Air Office's new-found interest in benefit analysis was rewarded with a brief report suggesting that the DOI slide show could not be given any technocratic credibility.[52] Although this finding might have induced the Air Office to reflect on the need for full scrubbing, the result was simply to deflate DOI's balloon. When the slide show was given a second showing on March 14 before a White House audience including cabinet members and high-level policymakers in the executive office, it once again received a hostile reception—with people from the DOE emphasizing that it was smelting, not coal burning, that would be the principal suspect in the case of Grand Canyon haze.[53] Since EPA could not deny this, the visibility argument on behalf of full scrubbing dropped out of sight during the last hectic months—only to reemerge at the final press conference when photographs of smoggy western wonders once again graced the walls as Administrator Costle announced the EPA decision.[54]

By mid-March, then, the clean air wing of the government had failed to sustain its opposition to a technocratic solution to the problem posed by Section 111. The time had come for the second half of the clean air–dirty coal coalition to make its presence felt.

Political Counterattack

Although the Planning Office's model was unsophisticated ecologically, the technocrats tried hard to take political realities seriously. The agency's computers modeled the impact of every proposed standard upon the coal mining industry, with special concern given the high-sulfur areas east of the Mississippi. Despite the anxieties of the high-sulfur coal lobby, the model consistently showed eastern coal production rising through 1995. The reasons for this pattern are not hard to understand. As we have seen, the bulk of existing power capacity will not be retired before 1995 and so old plants' demand for high-sulfur products will remain intact; similarly, the plants of the early and mid-1980s, regulated by the 1971 NSPS, will add their share to demand for eastern coals in the midsulfur range.[55] After the old plants finally retire in the twenty-first century, the plight of eastern coal can still be readily exaggerated. There are at least 30 *billion* tons of eastern coal that could be burned in compliance with the .55 standard—even under the most conservative assumptions about scrubbing removal and coal supplies.[56] Since at least half of this coal is commonly assumed to be economically mineable today, and only 540 *million* tons were mined in the East in 1975, we will be well into the new century before the East confronts a serious supply problem—at which point Americans may well have found a better way of desulfurizing coal, thereby permitting the use of the remaining 160 billion tons for exploitation in the East.[57] By any rational calculation, then, the problem of dislocation seems well within our control. Doubtless particular mines will close and particular mine owners will suffer, but workers will have a substantial period to adapt to the shift from higher to lower sulfur supplies. Indeed, the problem is of intergenerational dimension—the children of today's miners will work at new mines opened in response to changing demand patterns.

Yet, as their congressional activity indicated, the people threatened with short-run transition costs were more than able to protect their interests. Recalling the loss of northeastern markets to oil in the 1960s,[58] both eastern mine owners and the United Mine Workers saw their victory in the Congress of 1977 threatened by the technocrats of 1979. Requiring eastern coal burners to scrub their way to the 1.2-pound standard permitted the burning of practically all eastern coal, but lowering the ceiling to 0.55 would make all coals containing more than 3.7 pounds of sulfur unburnable at 85 percent scrubbing efficiency. Such a step, eastern coal spokesmen asserted, would "preclude" a substantial proportion of their coal from the market, causing serious damage to their industry.[59]

In making this "preclusion" argument, however, the coal lobby could no longer hope for the environmentalist support that had served it so well in the congressional struggle of 1977. For environmentalists, the new Section 111 had succeeded insofar as it induced the EPA to reduce the ceiling from 1.2 to 0.55. Although they preferred to go further and require full scrubbing in the West, they abandoned the 1.2 ceiling.[60] In contrast, the utilities welcomed their errant friends from the coal industry back into a dirty coal—dirty air alliance to resist the low ceiling. To the utilities, reducing the ceiling from 1.2 to 0.55 imposed intolerable cost burdens on coal burning. Moreover, utility spokesmen were happy to make the eastern mining complaint a part of their own rhetoric, decrying the preclusive effect of a lower ceiling on eastern coal.[61] Although the formal comment period had closed in January, the EPA's low ceiling proposal had only matured more recently. The agency responded to the coal-utility protest by holding a public meeting on April 5, at which the National Coal Association (NCA) presented an estimate that a 0.55 limit would preclude from 75 percent to 100 percent of the coal produced in five eastern regions.[62]

Unfortunately for the NCA, its definition of mining regions had been transparently devised to exclude major eastern zones of low-sulfur coal; hence its figures patently exaggerated the low ceiling's impact.[63] What is more, the NCA's entire conceptual exercise was radically misconceived. Measuring preclusion in percentage terms ignores the fact that, given the East's rich reserves, even a small percentage of available coal will keep miners fully occupied for decades; thus 1978 eastern production was 0.17 percent of total reserves.[64] As long as we assume that the next generation's pollution control technology will be superior to the scrubbers of today, it seems fair to leave the higher sulfur coal to tomorrow's treatment.

We do not mean to suggest that the NCA presentation fooled anybody. To the contrary, the inadequacies of the preclusion analysis were well recognized by the EPA bureaucracy.[65] Nonetheless, the agency was not in a position to give the NCA's special pleading the treatment it deserved. First, the congressional action of 1977—whatever its proper interpretation—was widely perceived as mandating a concern for "locally available fuels" as much as a desire for clean air. Second, and even more important, the agency had consistently behaved as if a new NSPS standard could be framed without a careful study of the benefits it would yield. When faced with the inevitable cries of discomfort from cost-bearers, the EPA could not defend its proposals by explaining why the new costs were worth bearing. Instead of countering the preclusion issue by pointing to the environmental benefits of a low ceiling, the EPA found the issue being defined in terms of political cost-benefit. As far as the coal industry was concerned, the EPA was breaking a promise explicitly made in the legislative history. And the industry would take energetic steps to bring the technocrats back into line with its understanding of congressional intention.

As April moved on, the big political guns moved in. Senate Majority Leader Robert Byrd of West Virginia met with Costle and Eizenstat in the senator's office on April 23 and again at the White House on May 2.[66] After the first meeting, Costle indicated that the agency did not intend to rule out large portions of eastern or midwestern coal reserves.[67] Capitulation followed. Not only did the EPA retreat from the new 0.55-pound to the old 1.2-pound number;[68] it resolved technical ambiguities surrounding the "old" 1971 number in ways that would permit new plants to increase legal discharges by as much as 50 percent.[69] At this new level, only 3 percent of eastern coal would be precluded from the market, leaving the nation free to choose its 0.2 percent annual requirement from the remaining 97 percent of eastern coal supplies.[70]

The political whirlpool was propelling Administrator Costle backward in bureaucratic time to the first Air Office proposal of early 1978—that required everybody to use 90 percent scrubbing under a 1.2 ceiling. Yet too much had happened to make this giant step backward easy to manage. Not only were there reams of computer printout detailing the weaknesses of the Air Office approach, but powerful bureaucracies were committed to the printout. Within the agency, a full scrubbing decision made under external political pressure would demoralize the Planning Office. Outside the agency, the Department of Energy remained adamantly opposed to a proposal that would place a 4-billion-dollar annual burden on energy independence by 1995;[71] and finally, the technocrats in the executive office were willing to defend their RARG report at any final White House confrontation.

But no bureaucrat likes confrontation. At the same time the EPA was propitiating Senator Byrd, it also tried to do homage to the gods of economic rationality. Ten days after the formal National Coal Association presentation, an internal memorandum suggested the high-level discovery of a new face-saving solution to the NSPS problem.[72] Relying on sketchy cost data provided by its Office of Research and Development, the EPA aggressively began to promote a new technology called the "dry scrubber."[73] The dry scrubber had a huge regulatory implication. Preliminary research indicated that a dry scrubber could operate more cheaply than an old-fashioned wet scrubber as long as it was not required to eliminate more than 70 percent of a coal's sulfur content.[74] The dry scrubber thus provided a symbolically satisfying way of justifying partial scrubbing. After all, the guiding principle motivating the agency's approach to NSPS had been the idea of technology-forcing[75]— would it not be wrong, then, to insist on 90 percent scrubbing if it would "preclude" a promising new technology like the dry scrubber?

The computers were rushed into action: how much would be saved if coal burners could use dry scrubbers to reduce the sulfur in coal by 70 percent? To emphasize the environmental promise of the new technology, the proposed regulation would allow 70 percent scrubbing only on the condition that a power plant would meet a low ceiling of 0.6 pounds of SO_2 per MBTU rather than the 1.2 pounds tolerated for 90 percent wet scrubbing.[76] In this way, the bureaucratic forces for a low ceiling got half a loaf; and when the computer results came rolling in, they revealed that the costs of a dry scrubbing–wet scrubbing sliding scale were far lower than the standard full scrubbing option and almost as low as the 0.55 ceiling. Moreover, given the 0.6 emission limit tolerated for dry scrubbing, the new proposal also generated a lower SO_2 wasteload than its full scrubbing competitor.*

With computer printout like this,[77] Costle could go to the White House with new confidence. Resistance within the executive office quickly faded. After all, it is embarrassing for the White House staff to urge the president to overrule a cabinet-level officer on an issue within his agency's jurisdiction. Now that Costle had moved away from full scrubbing toward RARG, the White House staff had gained a sufficient victory. And so, when a meeting with the president was finally held in early May, only Energy Secretary Schlesinger announced his commitment to the RARG low-ceiling option—and even he, from the reports we have received, was not outspoken.

Although the dry scrubber solved the administrator's political problem, its relationship to sulfur removal is more problematic. At present, there are no dry scrubbers operating on a full-sized

---

* The following table reports EPA's estimates of the costs and benefits generated in 1995 by competing NSPS programs:

	Incremental Cost ($ billions/ year)	National Emissions (millions tons/year)	Emissions East of Mississippi (millions tons/year)
Full scrubbing option[a]	4.42	20.72	15.81
Dry scrubbing option[b]	3.25	20.45	15.24
.55 Ceiling[c]	2.97	20.57	15.05

[a] 1.2 ceiling with 90% scrubbing
[b] 1.2/0.6 ceiling/floor, with either 90% or 70% scrubbing required
[c] .55 ceiling with 33% scrubbing required
SOURCE: ICF IV, *supra* note 4, at 5.

Note that the .55 ceiling costs less and produces fewer emissions east of the Mississippi than the dry scrubbing option. (EPA eventually promulgated the latter option.) The full scrubbing option appears inferior in all respects to both of the other options.

power plant anywhere in the United States.[78] Some experts we have talked to liken present knowledge of dry scrubbing to the level of understanding of wet scrubbing that prevailed in the early 1970s—when costs were thought to be only one-half of what we now know them to be.[79] Whether or not the dry scrubber will suffer a similar fate—indeed, whether it will work reliably under field conditions—only time will tell.

We do not mean to condemn a high-level decision to reach deep within the innards of EPA for intelligent guesses about the future. Instead, our objection involves the kind of guesses high-level decision makers thought relevant to their policymaking predicament. If EPA had been required to justify its decision in terms of environmental benefits, Administrator Costle would have had to call on agency modelers specializing in the long-distance transport of sulfates rather than on agency experts in frontier technologies. Doubtless the models produced would have been imperfect, requiring innumerable guesses before the link between midwestern smokestacks and eastern ill health could be elucidated. Doubtless the coal companies and Senator Byrd would have protested. At least, though, they would have been openly arguing about the right questions: How serious is the risk to human health? Are eastern lakes being irreversibly damaged by acid rain? If we are to spend billions, should they be spent on new plants or old ones?

Although answers to these questions are inevitably controversial, at least such a controversy would force policymakers to consider the consequences of their actions in the world beyond Washington, D.C. By moving beyond the New Deal, our institutions have moved away from a conscientious effort to understand the environment.

AGENCY-FORCING AND THE ROLE OF COURTS

THE AGENCY-FORCING STATUTE

At present the EPA's NSPS decision is on appeal.[1] Our study supplies several guidelines for a court reviewing the administrator's decision.

At the center of our story is a distinctive legal creature that we have called the agency-forcing statute. Agency-forcing provides a means for removing an issue from the general run of agency discretion and guiding policy in a particular direction. At the same time, however, it signals congressional recognition that the issue requires the exercise of expert judgment that cannot be applied directly from Capitol Hill. Thus, rather than setting down regulatory policy in explicit statutory terms, the agency-forcing statute contemplates careful policy reappraisal by the agency before a congressional initiative gains the force of law. Our task here is to define a role for courts in appraising agency performance—one that permits judges to check the worst abuses generated by agency-forcing without presuming a policymaking competence that outstrips judicial capacities.

THE PRINCIPLE OF FULL INQUIRY

Reflect upon an obvious danger generated by agency-forcing. Simply put, it is the old Alphonse–Gaston problem: one player, Congress, enacts an agency-forcing statute with the expectation that the other player will subject a particular policy to hardheaded consideration. The second player, the agency, thinks that Congress has already made the policy judgment and narrowly confines its policy review. Each player allows the other to drop the ball: an important policy is adopted without the hard thinking that should be required of a sound lawmaking enterprise.

To avoid such breakdowns, the court must adapt a classic function of judicial review to the agency-forcing context. Traditionally, courts have been alive to the importance of scrutinizing administrative procedures to assure the full airing of policy options before an agency's decision.[2] Given agency-forcing statutes, however, this focus on agency procedures—although still important—is no longer sufficient. The judicial concern with a full policy airing must also inform the court's approach to problems that arise in the interpretation of the statute itself. Agency-forcing statutes should be read in the light of the *principle of full inquiry*—requiring the fullest possible agency investigation into competing policy approaches consistent with the text of the agency-forcing statute.[3]

To make this principle more precise, distinguish several kinds of statutory interventions within the agency-forcing category. The first and least aggressive statute simply forces the agency to focus decision-making attention on a particular question, the desirability of scrubbing for example, rather than on the countless other policy questions that might occupy its time.[4] In contrast to this *agenda-forcing* function, more assertive statutes aim to give a particular policy solution special salience in the agency's deliberations—call this *solution-forcing*.

Vocabulary taken from the law of evidence can help analyze these more ambitious efforts. The most modest statute might simply allocate the *burden of coming forward* without affecting the ultimate burden of persuasion.[5] For example, Congress might instruct the EPA to select the full scrubbing option if the agency cannot generate any plausible technocratic analyses that will better inform its deliberations. More active legislative involvement comes through a shift in the *burden of persuasion*.[6] The agency might be instructed to accept full scrubbing unless it is persuaded that the policy is unwise. Finally, the agency-forcing statute may alter the *burden of proof*.[7] It may specify that the EPA shall impose scrubbing unless the agency is convinced beyond all reasonable doubt that scrubbing is unwise.

Although these distinctions are familiar, they have not customarily been invoked in the interpretation of agency-forcing statutes—because the distinctive character of these statutes has been insufficiently appreciated. Having imported these familiar concepts into a new area, it is possible to define the principle of full inquiry in judicially manageable terms. In a nutshell, courts should not be quick to read the more aggressive forms of agency-forcing into ambiguous statutory language. A congressional instruction that may be plausibly interpreted as an exercise in *agenda*-forcing should not be escalated into a form of *solution*-forcing. A statute that may be read as allocating the burden of coming forward should not be blown up into a congressional judgment about the burden of persuasion. And so forth. Easy escalation runs the risk of the Alphonse–Gaston mistake.

In contrast, the risk of error involved in an interpretative policy of de-escalation is far smaller. After all, in passing the agency-forcing statute, Congress expressed doubts as to its own capacity to impose a decisive solution on the problem at hand. Even if, by hypothesis, the agency has given the congressional initiative greater scrutiny than Congress intended, the fact that the proposal does not survive hardheaded examination is hardly insignificant. Indeed, when apprised of the agency's finding, Congress may well applaud the action retrospectively.[8] Given the conditions of information overload that prevail on Capitol Hill, agency resistance may well induce Congress to focus more seriously on the hard facts that lie beneath the pretty symbols.

A similar perception is at the core of Judge Leventhal's landmark decision in *International Harvester v. Ruckelshaus*.[9] Congress had ordered automakers to reduce automobile emissions in 1975 but allowed the administrator to defer this legislative solution by one year if the carmakers satisfied four stringent criteria.[10] The administrator denied the companies' request for the year's delay, relying on a restrictive view of the range of inquiry that the statute permitted him. This was unacceptable to Judge Leventhal, who detected the risk that a decision with serious effects on the nation's economy was being generated by an Alphonse–Gaston process. Rather than run this risk, Leventhal interpreted the statute as requiring broader administrative inquiry.[11] The principle of full inquiry develops Leventhal's insight into a more general guideline—preventing the agency from arbitrarily constricting the range of policy choices left open by the statutory text.

In the case of the scrubbing controversy, the principle of full inquiry requires a strong remand to the EPA for further consideration. Rather than reading Section 111 as an agenda-forcing provision, the EPA read it as a strong form of solution-forcing. At no point did the agency seriously consider the possibility that the costs of universal scrubbing were intolerably high. Instead it constructed a *conclusive* presumption in favor of requiring *all* new plants to scrub and only considered whether some plants should be permitted to scrub less than others. Yet, as we have shown, a fair reading of the statutory text did not demand this narrow focus. Not only was the administrator explicitly required to consider costs, but the statutory text gives no indication of a solution-forcing intention.[12] The new amendment to Section 111 merely changes the *form* of EPA regulation—requiring the agency to set percentage treatment requirements that may vary with different kinds of power plants. Although this change in regulatory form certainly counts as an agenda-forcing statute, it violates the principle of full consideration to promote it into the strongest form of solution-forcing, authorizing the agency to exclude low-sulfur coal strategies arbitrarily from its analysis of relevant policy options.

THE PRINCIPLE OF TEXTUAL PRIORITY

But was the EPA not justified in looking beyond the statutory text to the legislative history—pregnant as it is with the rhetoric of the eastern coal lobby? In rejecting this move, we do not wish to deny legislative history a proper place in the armory of statutory interpretation. We only insist that its place be defined after due consideration of the peculiar abuses attendant upon the passage of an agency-forcing statute.

After all, agency-forcing occurs only when congressional interests are insufficiently powerful to enact their favored policy into explicit statutory language. There can, for example, be no doubt that the clean air–dirty coal coalition would have vastly preferred a single statutory phrase commanding full scrubbing to a ream of legislative history in praise of "locally available coal." The reason they settled for less is that they feared they would lose if they went for broke and forced a statutory showdown on NSPS. Such fears seem rational when the 45–44 Senate victory of Section 125 is recalled. Here was a section that explicitly authorized the use of scrubbing for the protection of eastern coal; yet, while it was hedged with all sorts of checks, it gained the narrowest of victories. This fact speaks louder than ten thousand words of legislative history in praise of a full scrubbing proposal that dwarfs the bail-out operation Congress so hesitantly endorsed. If it was so hard for eastern coal to gain the uncertain relief promised by Section 125, why should a court presume that Congress "intends" the expenditure of tens of billions of dollars when this intention has not yet achieved clear expression in the text of Section 111?

But it would be a mistake to place too much weight on the contextual evidence provided by Section 125. The pure theory of the agency-forcing statute itself suggests judicial caution in the use of legislative history. When considered as a distinctive lawmaking genre, the agency-forcing statute seems peculiarly prone to hyperbolic treatment in committee reports. If an interest group is sufficiently powerful to lobby an agency-forcing statute through Congress, it will be strong enough to gain access to the committee reports as well. And, given the weak controls on the content of legislative history, a group with access can stuff the reports with language far stronger than anything that would survive on the surface of a statute. It is only by insisting on explicit statutory language that courts can assure themselves that the committee reports represent more than a successful effort by a handful of insiders to exploit the overloaded congressional docket.[13] Over time, the regular application of this *principle of textual priority* will bring home to the staff and lobbyists on Capitol Hill the realization that there is only one way to force an agency to do their bidding—and that is to engage in the full debate traditionally associated with explicit statutory amendment.

Textual priority is especially important in the interpretation of a large and complex statute like the Clean Air Act. Such statutes have their own distinctive architecture—with a few basic principles motivating the often intricate ground plan. Although particular agency-forcing provisions may sometimes explicitly require regulators to depart from basic statutory purposes, courts should not destroy the ground plan on the basis of subterranean legislative history. To endow our point with the dignity of a platitude: *the basic purpose of the Clean Air Act is to clean the air*.[14] Reading the "new" Section 111 as an agenda-forcing statute will serve this basic purpose by encouraging EPA to invest more of its resources in sensitive long-range planning. Reading Section 111 as if it were a solution-forcing statute will subvert this basic purpose—instead of cleaning the air, such a reading requires the administrator to subsidize the producers of polluting products. When confronted with a clear affirmation of a policy of regional protectionism, as in Section 125, both agencies and courts may have no choice but to give way.[15] In the absence of a clear statutory text, however, neither agency nor court should read Section 111 as encouraging a massive bail-out for sectional interests.

THE COORDINATION PRINCIPLE

Agency-forcing not only raises distinctive problems in the coordination of congressional intervention and agency response; it also imposes special strains upon the executive branch. Forcing an issue onto one agency's agenda will generate ripple effects as other agencies react to the agency forced into action. These responses, in turn, impose pressures on the White House to help shape the congressional initiative into a policy that gives due regard to the missions of executive agencies other than the one initially targeted by Congress. As a consequence, an agency may find itself awkwardly placed between a congressional directive, on the one hand, and an executive mandate, on the other.

Such contretemps can obviously raise constitutional issues.[16] Yet before a court forces a confrontation to that level, it should search for a plausible statutory interpretation that renders it unnecessary. This search may prove especially rewarding in the agency-forcing context. After all, in refraining from specifying a final solution, Congress itself has recognized the need for further agency study. If agency analysis reveals a sharp conflict with the goals of other agencies—goals whose importance Congress has endorsed in other statutes—courts should not be quick to say that the agency-forcing statute forbids policymakers from taking these competing goals into account. Otherwise, a narrowly focused and tentative statutory directive may undermine policies that Congress may value more highly if it had fully appreciated the broader consequences of its agency-forcing mandate. This *principle of coordination* is particularly appropriate when reading statutes like the Clean Air Act that explicitly require interagency cooperation.[17]

On some occasions, the coordination principle might require an agency to undertake the bureaucratically difficult task of considering a legislative goal far removed from its ordinary mission.[18] In the present case, however, there is no need for a court to read the statute in a way that imposes such an onerous burden. The White House staff did not try to induce the agency to compromise its environmental mission by giving affirmative weight to the nation's energy goals. Instead, the executive's RARG report focused on the agency's failure to bring its *environmental* expertise to the problem at hand. In essence, RARG was saying that before EPA imposed a multibillion-dollar burden on the congressionally approved goal of energy independence, it should, at the very least, show how this would yield a better environment. Rather than deploy its expertise to coordinate policy in this minimal fashion, the EPA responded by reading Section 111 in a way that made environmental sophistication unnecessary. Given the principle of full inquiry, EPA's narrow view should be rejected by the courts in any event. Yet the RARG report should be taken as an indication that the EPA's narrow response to the agency-forcing statute generated real problems of executive coordination; and that, when the issue returns to the EPA, it should make a minimal effort to coordinate its policy with other agencies by assuring that the sacrifice of their competing goals will generate *real* environmental gains. More particularly, the court should encourage the EPA to make use of its latent expertise and consider the possibility that there are cheaper, quicker, and more certain ways of responding to the only environmental problem that would remotely justify its costly drag on energy independence—the risks that sulfates, not sulfur dioxides, raise to eastern health and ecology as well as to western visibility.[19] It is only in this way that EPA can discharge its explicit statutory obligation to take costs "into account" before imposing billion-dollar burdens on the American people in the name of clean air.

BEYOND FORMALISTIC REVIEW

There is a final ground for judicial intervention. When viewed from a formalist perspective, it is easy to find procedural flaws in the complex story we have recounted. The agency's heavy reliance on dry scrubbing was hardly to be anticipated from a careful reading of the September notice that the EPA published in the *Federal Register* to initiate its rule-making proceeding. And a host of agency outsiders—ranging from Senator Byrd to Secretary Andrus—made their presence felt within the EPA after the formal close of the comment period on January 15, 1979.

Not that the agency lacked procedural sensitivity in responding to the evolving policy dispute. As the Planning Office's advocacy of a low emission ceiling came to the fore, EPA published a supplementary notice in the *Federal Register* that detailed the most recent modeling results.[20] Later, as the eastern coal industry's counteroffensive became salient, the EPA held a public meeting on April 5, 1979, at which all sides took the opportunity to comment upon the National Coal Association's concerns. A similar concern is apparent in the agency's dealings with outsiders. All such contacts are scrupulously recorded in the docket—with names and dates provided even when the outsiders are merely representatives of other executive branch agencies.

As always, however, it is possible to say that the EPA did not go far enough in proliferating the forms of due process. When dry scrubbing gained prominence at the last moment, EPA did not stop to invite yet another round of comments in the *Federal Register*. Similarly, the agency docket does not contain elaborate summaries of the discussions held between outsiders and agency officials. And doubtless there are even more stringent formalisms that could have been imposed on the process we described. Nonetheless, we believe that judges should resist such formalistic temptations and sustain the agency's procedural handling of the case. Forcing the agency to design additional due process mechanisms will threaten the integrity of more important values.

To glimpse a first danger, recall the relative innocence with which the EPA viewed the scrubbing problem at the time it filed its September notice in the *Federal Register*. The agency had only begun to redeem a decade's analytic neglect. The computers were humming but the results had not yet been assimilated. Now imagine that Administrator Costle had known, in September, that he would be obliged to go through another round of note-and-comment rule making if the September–January round generated some new insights and proposals. Would he have authorized the Planning Office to proceed with its modeling activity? There is always one way to assure that nothing new will come out of a rule making proceeding—and that is to refuse to invest analytic resources in the learning process. Agency learning should not be penalized when it uncovers something new. There are enough temptations to ignore economic and ecological realities without the law making learning more costly.

The threat to agency learning is only one half of the price of undue formalism. Paradoxically, formalism will also endanger the very values of fair notice and effective participation it seeks to assure. At present, EPA officials often meet with outsiders at both technical and policymaking levels, getting new ideas and informally gauging responses to possible regulatory initiatives. Nor is this interchange restricted to a privileged few—a reader of the weekly BNA *Environmental Reporter* would have no difficulty following the main course of the decision reported here. Yet a formalistic approach to outside intervention would jeopardize this rich set of informal contacts, destroying a vital supplement to the inevitably inadequate publications in the *Federal Register*. The National Coal Association, for example, was able to intervene effectively in the NSPS decision because of the informal notice and comment mechanism at EPA. Rather than hiding behind the formal closing of the comment period and refusing to discuss the impact of its chosen standard on an important industry, EPA allowed the NCA, in the presence of its opponents, to advance its coal preclusion argument. An agency that responds in this creative fashion should not be subject to Monday morning quarterbacking from the bench.

We take a similar view of even the most dramatic forms of outside intervention revealed by our study. Although we believe it was wrong of the agency to cave in before Senator Byrd, this is not because a few officials actually spoke to the senator in the White House.[21] A similar meeting could have been arranged by inviting EPA staff to testify before the Senate Committee on Environment and Public Works, which is chaired by the other senator from West Virginia.[22] In any event, the administrator would undoubtedly know the opinion of the Senate majority leader on an issue of such importance to his home state.

Moreover, as long as one concedes the relevance of the legislative history, Senator Byrd had a good point: was the EPA, in proposing a low ceiling, doing all it could in the name of "locally available coal"? If the question is legally relevant, we can think of nobody more appropriate to raise it than the senator from West Virginia.* Because the senator's meetings were entered in the docket, everyone had notice of their existence.[23]

The problem with Byrd's intervention was not procedural but substantive. It was simply another battle in the war begun by the clean air–dirty coal lobby in 1976 to escalate its carefully drafted legislative history into law. The agency was wrong to listen to Byrd for the same reason it was wrong to allow the rhetoric of the House committee report to narrow its scrubbing inquiry. The cave-in before a powerful senator only dramatizes the danger of moving beyond the text of an agency-forcing statute.[24]

Rather than grasp at procedural straws, the only appropriate way for the court to intervene is through *substantive* construction of the agency-forcing statute. The main concern should not be with the senator's protest but with the agency's narrow conception of its mission that made capitulation seem so legitimate. Section 111 did not force a solution on the EPA. It only obliged the agency to give the scrubber a high priority on its decision-making agenda and required the EPA to use its expertise to see whether the scrubber made environmental sense given its heavy costs. Douglas Costle should not be criticized because he chose to

---

* Except perhaps the senators from Ohio.

speak with members of the body that supervised his policy-making. Instead, the agency should be criticized for failing to relate scrubbing to what the act most plainly requires: cleaning the air in a cost-effective way that yields real environmental benefits.

We should be clear about what we have *not* been saying. At no point have we suggested that the courts should take primary responsibility for the development of substantive policy under the Clean Air Act. The objective of judicial review remains as it was under the classical New Deal conception—to assure a full and focused airing of plausible policy options before officials make decisions of consequence. The move beyond the New Deal agency requires, however, that courts delicately calibrate Congress-agency relationships to prevent mindless decision making.[25] Careful employment of the three maxims of full inquiry, textual priority, and coordination will avoid needless and costly disappointments that may otherwise result from the effort to venture beyond New Deal ideals.

Applied to the scrubbing controversy, these principles demand a strong remand to the administrator. On reconsideration, he should not take the asserted threat to eastern coal as a sufficient reason for requiring scrubbing. Nor should he ignore RARG's insistence that before the agency imposes the multibillion-dollar cost of forced scrubbing on the American people, it should point to environmental benefits that remotely justify the burden.

8

REFORM

LEARNING FROM EXPERIENCE?

But there is more to administrative law than judicial review
Even if the courts do save us from a particularly mindless decision by sensitive statutory construction, larger doubts remain. It
is a sign of underlying weakness that policy coordination between Congress and the EPA depends on acts of judicial states-
manship. While courts may sometimes be equal to the challenge
they will often fall short. The ultimate question raised by our
study is the wisdom of the Clean Air Act's effort to move beyond
the New Deal agency.

   The story we have told is an ironic tribute to discarded New
Deal virtues. The Clean Air Act discarded the idea that concrete
policy decisions should be made in a decision-making forum insulated from direct congressional intervention. Yet the legitimacy gains expected from direct congressional decision failed to
materialize. Instead of a solution responsive to the evolving will
of a national majority, congressional intervention mixed clean
air symbols and dirty coal self-interest in a way that invites cyni-
cism about democratic self-government. Similarly, the Clean Air
Act rejected the idea that responsible policymaking requires a
sober sifting of the facts so that regulatory means are carefully
crafted to achieve desired ends. Yet the proud legislative
affirmation of the symbol of technology-forcing only serves to
raise doubts about the extent to which technology can answer
our ecological predicament. Such unhappy results force a re-

116

treat to basic premises: is it naive to imagine that a different institutional design could have avoided the fiasco we have described? If administrative design *can* make a difference, what institutional setup might have channeled decision making in a more fruitful direction?

THE IMPORTANCE OF INSTITUTIONAL DESIGN

It is easy to read this study as yet another despairing commentary on the capacity of government to improve our lives. On this view, the political forces behind forced scrubbing were well-nigh irresistible. It is simply naive to imagine that subtle changes in institutional design could have altered the unhappy outcome. Nothing short of a radical transformation of Congress, or a return of air pollution policy to the states, would have had an impact on the process by which the Clean Air Act was transformed to serve the purposes of dirty coal. Such a despairing view, it should be emphasized, need not lead to the advocacy of drastic changes in the status quo. If faced with a choice between the bastardized Clean Air Act of today or the pathetically weak legal regime prevailing before the first Earth Day, no doubt we would settle for what we have. We deny, however, that a sober sense of political reality leads to such grim alternatives.

The clean air–dirty coal coalition was *not* an irresistible political force but a fragile coalition depending for its survival on peculiar conditions fostered by the congressional move beyond the New Deal. Reflect on the delicate strategic position of the eastern coal lobby.[1] As far as it was concerned, joining forces with the environmentalists was an act of desperation. Rather than push for forced scrubbing, eastern coal would have vastly preferred to weaken the 1.2 NSPS ceiling, thereby permitting new plants to burn dirty coal *without* scrubbing. The only reason eastern coal interests abandoned their long-standing alliance with the public utilities was their (sensible) belief that their second-best strategy was far more likely to succeed under the peculiar conditions generated by an overloaded congressional docket.

However, if eastern coal were forced to pursue its interests within a different institutional framework, it would have lost all incentive to join forces with the clean air lobby. Imagine, for example, that the NSPS decision were in the hands of an insulated and expert pollution control agency concerned with the cost-effective pursuit of health-related targets established by the Congress.* Rather than exploit congressional ignorance, the coal industry would then have been obliged to appear before hearing examiners who had both the time and the bureaucratic incentives to put new plants in environmental context. Given this new institutional framework, eastern coal would not have maximized its self-interest by making the doubtful instrumental case for forced scrubbing. Instead, it would have best served itself by maintaining its traditional alliance with the utility industry—investing in a technocratic effort to convince the agency that the 1.2 standard was unjustifiably stringent in light of the health objectives set by Congress. Although this strategy might not have been successful, the critical question is whether, *ex ante*, its value was higher than any other strategy that might be pursued before our hypothetical agency. And as long as that was the question, there can be no reasonable doubt as to eastern coal's answer: since an expert cost-minimizing agency would so obviously reject forced scrubbing, eastern coal's only hope was to join with the utilities in an effort to convince the agency that it had overestimated the harmful impact of new plant emissions on the nation's health. Institutional design, then, *can* make a difference: eastern coal would have abandoned its coalition with

---

* Not that we presently know how to design an expert agency of the kind hypothesized. Nor can we hope, in the present case study, to confront the fundamental questions of institutional design that must be resolved before our hypothetical agency could seem a credible reality. All this book can do is emphasize the need for such work if we are to avoid the endless repetition of the scrubbing scenario.

environmentalists to rejoin its friends in the utility industry if NSPS had been fought out before an expert and cost-conscious agency insulated from direct congressional intervention.

The strategic significance of institutional design is hardly an idiosyncratic feature of our peculiar case study. It is the product of two economic factors that will arise in a broad range of environmental controversies. To isolate the first factor, consider that most pollution is a *joint product* of a vertically integrated stream of economic activities—thus, smokestack emissions are a product of *both* coal mining *and* power generation; auto emissions, a product of *both* gasoline refining *and* the internal combustion engine; and so forth. This means that policymakers can cut back on pollution by applying pressure at a number of different stages in the stream of production—the clean-up cost can be borne, in different proportions, by coal mining or power generating, oil refining or car manufacturing, depending on the particular control measures adopted. This fact would be of little importance if all stages in the stream of production were owned by a group of vertically integrated companies. If a single firm owned *both* coal reserves and power plants, it would have an incentive to choose the cheapest way of reducing its final discharge, regardless of the corporate division that bore the costs of cleanup. If it cost the company's coal division an extra million dollars to produce low-sulfur coal, while it cost the power plant division two million dollars to scrub, the integrated firm would have an incentive to maximize profits by choosing the low-sulfur option.

It is at this point, however, that a second basic economic fact enters the strategic picture: *incomplete vertical integration*. In other words, all the relevant stages in the production process are not owned by the same profit-making units: the Peabody Coal Company is not Edison Electric; Exxon is not General Motors. While this may be a very good thing from other points of view, incomplete vertical integration fuels the strategic instability exemplified in our present case. Even if pollution will be reduced most

cheaply by pursuing a low-sulfur coal strategy, profit-maxi-
mizing coal companies may try to shift the clean-up burden to
some other stage of the production process. Thus, if the costs
to Coal, Inc. would be one million dollars to achieve a pollution
cutback, it may be willing to spend almost that amount to force
Power, Inc. to spend two million dollars in its stead.

Moreover, if Coal, Inc. makes an overture to the clean air
lobby, it may well find a sympathetic audience for its campaign
to impose the clean-up burden on the utilities. After all, if the
clean air lobby remains unsympathetic to the goals of both Coal,
Inc. and Power, Inc., its political position will be inferior to the
one it would occupy if it could split the energy coalition. The
clean air lobby has an incentive to exploit incomplete vertical
integration, despite the resulting increased overall costs to
society.[2]

Yet even at its most tempting, a clean air–dirty coal coalition is
full of dangers for Coal, Inc. The clean air lobby might manipulate the coalition so successfully that the law-making process
generates far more stringent cut-back requirements than would
have resulted if the lobby had been obliged to stand alone
against a united coal-power coalition. And if this happens, Coal,
Inc. may end up paying more than a million dollars even if it
manages to deflect a larger *proportion* of the total clean-up cost
onto Power, Inc. Since the total clean-up burden may now be
larger than it would have been otherwise, a smaller fraction of
that larger amount may well add up to more than a million dollars. In short, incomplete vertical integration imposes strategic
dilemmas on firms at different stages of the production process.
On the one hand, a group of producers may pursue a *cost-
externalization* strategy—joining forces with environmentalists
in an effort to impose costs on other production stages but at the
risk of increasing the total clean-up burden. On the other hand,
they may pursue a *cost-minimization* strategy—joining forces
with their fellow producers in a battle against the environmentalists to weaken the standards.

And, as each firm considers its strategic options, the institutional framework will often shift the balance of advantage from cost-externalization to cost-minimization, or vice versa. If Congress had consigned the NSPS issue to our hypothetical expert agency concerned with the cost-effective pursuit of health-related targets, firms at different levels of the production process would have had little to gain from the cost-externalization strategy. Rather than join with their environmentalist adversaries in attempting an end run around the agency, it would generally seem more sensible for them to join with firms at other stages in the production process in an effort to minimize overall compliance costs.* Indeed, an initial success in insulating the agency from Congress will feed on itself—the more firms believe that Congress will respect the agency's insulation, the less incentive they have to form bizarre cost-externalizing coalitions in an effort to overturn cost minimizing decisions, and the fewer opportunities Congress will have to move beyond the New Deal in the unhappy way dramatized by our study.

ENDS AND MEANS IN ENVIRONMENTAL LAW

Just as our story need not end in despair, so too it should not

* We can think of one important situation where groups of producers might pursue the cost-externalization strategy even when confronting our hypothetical expert agency. In this troublesome situation, the cost-minimizing share of clean-up costs borne by each production stage shifts dramatically with a change in the overall discharge limit. Imagine, for example, that coal washing at the mine were the cheapest way of achieving an emission limitation of 2.0 pounds per MBTU or higher. If, however, the agency selected a lower standard, then it would be cheapest to ship *unwashed* coal to the power plants and require Power, Inc. to scrub. If this cost structure prevailed (which it does not), then Coal, Inc. might still have an incentive to join with the clean air lobby in a campaign for stringent emission standards. To avoid the coalition in such cases, the agency must adopt a complex tax-subsidy approach that requires Coal, Inc. to bear a share of *Power, Inc.*'s clean-up costs if the power company were found to be the cheapest cost-avoider. For some further (inadequate) reflections on the link between tax-subsidy schemes and political activity, see B. ACKERMAN, PRIVATE PROPERTY AND THE CONSTITUTION 54–57 (1977).

end in nostalgia. Although the New Deal tradition *does* have much to teach us as we struggle to define a sound environmental policy, Congress was not wrong in finding it wanting at the time of the legal revolution of 1970. We are not inclined to dismiss the fear of EPA "capture" by polluting interests that so obviously motivated the effort to move beyond New Deal models.[3] While the theory of "agency capture" requires more work than it has been given,[4] the experience of an entire generation cautioned legal activists against contenting themselves with a statute that, Polonius-like, told the infant EPA "to protect the environment while protecting the economy." In 1970, the country was ready for an aggressive effort to reduce pollution. Environmentalists were right in trying to ensure that this outpouring of public sentiment would not be sabotaged by administrative passivity. The idea of agency-forcing was a legal innovation responsive to this need.

Where the 1970 act went wrong is not in its basic diagnosis but in the particular cure it prescribed for the disease. Agency inaction can be cured in two different ways. The first, and more primitive, may be called "means-oriented" agency-forcing. Here the agency is told precisely which regulatory means should be used to reach the congressional objective. This approach, of course, informs congressional policy toward new coal burners. High-technology solutions are given special prominence they would not deserve in a sensitive exercise in means-end rationality. Under the second form of agency-forcing, Congress requires the agency to define its *ends* aggressively and challenges the agency to select a course that promises to reach its goals effectively. This approach was taken in Sections 108 and 109 of the act—requiring the administrator to establish ambient air quality standards "to protect the public health" and setting 1977 as the date by which the United States was to reach these initial clean air objectives. Unlike its competitor, "ends-oriented" agency-forcing does not require Congress to indulge in instrumental judgments beyond its capacity. Instead, it generates a process by which the ultimate aims of environmental policy can be clarified over time.

Consider the way Congress's ends-oriented exercise of 1970 shaped the debate of 1976 and 1977. Pursuant to directive, the administrator specified air quality objectives and tried to prod the states into speedy compliance. By 1977, this effort had generated some simple, but important, lessons. First, the original congressional effort to specify objectives had failed to consider the peculiar values at stake in the clean air regions. Thus, 1977 became an occasion for *clarifying the nature of new objectives* for environmental policy. Second, reaching the old objectives turned out to be more costly and difficult than imagined. And so 1977 became a time for *reappraising old objectives* and deciding how forcefully to pursue them in the future. A reader of the legislative history cannot but be impressed by the extent to which Congress lived up to these responsibilities in 1977.[5] While we may disagree with Congress's particular conclusions, at least congressmen were alive to the basic value trade-offs at stake. Contrast this success with the sorry story we have told. The act's 1970 exercise in means-forcing set the stage in 1977 for the effort by the dirty coal–clean air coalition to exploit the technical incompetence of Congress to advance ends that are peripheral to the main goals of environmental law.

There is no undoing the past, but surely we can hope in the future to come to terms with the complexities of agency-forcing—building upon the ends-oriented achievements of the past while trying to ameliorate the harm generated by the means-oriented approach. At a minimum, this means integrating NSPS requirements into the basic framework of instrumental rationality established by the act. From an ecological point of view, a pound of pollution is a pound of pollution, regardless of the age of the polluter. Legal regulation of new plants should not be allowed to generate a fantasy world where this basic ecological fact is obscured from view. While a long-term strategy for new plants is important, a short-term strategy should allow polluters to trade costly high technology in new plants in exchange for pollution cutbacks coming out of old plants. Although this offset principle has already gained some statutory recognition,[6] courts should support EPA efforts to make the offset program operational and to broaden its application to new plants.[7] By emphasizing the trade-off between old and new plants, the EPA will save money while educating Congress in the hazards of means-oriented agency-forcing.

No less important is the effort to build upon the act's valuable achievements in ends-oriented agency-forcing. We should learn to look with a critical eye on instructions, like those found in Section 109, to set clean air targets in a way that "protect[s] the public health" while "allowing an adequate margin of safety. . . ." Such vague formulas represent the unhappy legacy of a New Deal past that too readily delegated basic value choices to bureaucratic "experts." Instead of speaking vaguely of the "public health," the task is to define clear and operational goals through the democratic process and then challenge the experts to meet these goals in a fair and efficient fashion by a specified date.[8]

Imagine, for example, an amendment that would have required the administrator to "achieve ambient air quality improvements that promised to add at least 25,000 years to the life expectancies of the American people by 1984." Such a statute would have forced agency action of a very different kind from that described in our case study. To begin, the statute would have forced agency policymakers to ask themselves which of countless possible regulatory initiatives could plausibly be related to the 25,000 life-year goal. As soon as the matter is framed in this way, decision makers would become acutely aware of the state of our ignorance as to the health effects of air pollution. Yet the fact that the statute would require the agency to engage in high-visibility guesswork is *not* an argument against ends-oriented agency-forcing. To the contrary, this is its greatest institutional strength, for two reasons. First, with the need for guessing out in the open, it may no longer be so embarrassing to admit that some guesses have turned out wrong. To take an example from our case study, it is now conventional wisdom that SO_4, not SO_2, should be the sulfur oxide of primary concern.[9] Nonetheless, given the force of bureaucratic inertia in support of the current EPA focus on SO_2,[10] it is not clear when—if ever—this recognition will be reflected in the ambient standards that purportedly protect "the public health." In contrast, new bureaucratic incentives would emerge under an ends-specifying strategy of the kind hypothesized. Here, if there were reason to argue that SO_4 control would help the agency meet its 25,000 life-year target, there would be new incentives to overcome agency inertia, change the pollution indicator, and design a program that would permit the presentation of a credible progress report to Congress in 1984.

Second, high-visibility guesswork would require policymakers to take the question of research priorities far more seriously than they do at present. Whatever may be said about the myth of administrative expertise in other areas, the EPA today has the capacity to launch an enormously useful research enterprise. And there is no better way to interest policymakers in the direction of research than by linking their reputations to the data that is gathered and analyzed.

Not that ends-oriented agency-forcing is offered as a panacea. The present state of model building, even data monitoring, permits a wide range for agency discretion. And agencies will use this discretion in ways that will generate good report cards when the time comes for congressional oversight.[11] It is a good bet that if the 25,000 life-year goal had been announced in 1977, the EPA would have found a way to have its computers declare that its policies had saved far more than 25,000 years in 1984. While this tendency toward over-optimism cannot be eradicated, steps may be taken to counteract it. Before going to Congress, the agency's septennial progress report might be submitted to several ad hoc committees of experts—each group selected from a *different* place on the value spectrum generated by environmental protection. If provided with the budget needed for a serious critique of agency premises and predictions, these panels could be expected to expose the darker shadows ignored by the agency's predictably rosy progress report. Even if many doubtful premises remained unexplored, our study suggests that it is better to give an optimistic answer to the *right* question—what is happening to the air?—than to engage in a desperate attempt to answer the *wrong* question—how well do scrubbers work?

Although ends-forcing promises, over time, to improve the quality of EPA decision making, perhaps its most important contribution will be to the quality of future congressional deliberations. If the Congress of 1977 had set a 25,000 life-year target, the debate of 1984 would naturally focus on the ends that had been specified. Agency self-interest would motivate an EPA demand for expanded operational goals—framed, in keeping with the ends-forcing orientation, in terms of physical discomfort and aesthetic affront. And groups representing energy users would band together to weaken the goals. Rather than invite bizarre coalitions between clean air and dirty coal, the focus on operational goals would invite democratic tests of opinion between pro- and anti-environmental forces.

It is not too utopian to imagine this process leading to the further refinement of environmental objectives. Instead of directly assaulting the 25,000 life-year objective, the pro-energy forces might respond by proposing a new amendment: "Save 25,000 more life years by 1991 *subject to the constraint that polluters spend no more than 25 billion dollars a year in control costs*." Given this overall budget constraint,[12] the EPA could then allocate its budgetary authority among a variety of regulatory projects that sought to achieve its 25,000 life-year goal. To do this, the regulatory cost of different options would have to be calculated; and once again, the computers could be expected to generate numbers that gave the agency the benefit of the doubt. The critical question, however, is not whether an exercise in agency-forcing leads to the "correct" decision in each case,* but whether it invites policymakers from both Congress and the EPA to direct their attention to matters that exploit their comparative institutional advantages.

Once again, we do not want to paint too rosy a picture. Doubtless the Senate of 1984 will contain two advocates of the West Virginia coal mines, and they may try to exact an understanding about scrubbing as a condition for aiding the campaign for a pro-environmental specification of EPA ends. Nonetheless, a shift in the form of agency-forcing would focus congressional debate on the issues Congress is most competent to handle. Rather than indulge in a harried pseudotechnocratic judgment about the scrubber, Congress would be encouraged to concentrate on basic questions. Are we doing too much or too little in the way of environmental protection? Do our old goals still make sense or do other objectives merit greater emphasis in the years ahead? How well is the EPA really doing in meeting its operational goals and setting new objectives that seem sensible?

It is not enough, however, for Congress to monitor agency performance at regular intervals and set operational goals for the next septennium. Beyond the next decade, there is the next generation. And our case study gives eloquent testimony concerning the ease with which long-range planning can be sacrificed to meet middle-range goals set by Congress. To discharge our minimal obligation to future generations, we must assure that they will have a better understanding of the world we have bequeathed them. In pursuit of this end, Congress must massively increase funds for monitoring and modeling the environment. And at every septennial review, a special branch of EPA charged with long-run planning should be invited to focus attention upon the most salient policy redirections suggested by

---

* Not that this is a matter of indifference. We would, however, leave it to the courts, aided by RARG-type executive institutions, to expose and invalidate the worst abuses of expertise as "arbitrary or capricious." See pp. 6–7, *supra*.

the direction of scientific research.[13] In terms of this study, what seems most urgent now is a redefinition of the sulfur oxide problem to emphasize sulfates, rather than sulfur dioxides, and the organization of a set of regional institutions capable of handling the problem of long-range transport.[14] But it is always possible that ten years more of research will make this now-conventional wisdom as dated as the EPA's fixation on SO_2 seems today. What is required is a branch of the EPA that does not see obsolescence as a bureaucratic threat but as a source of institutional power.

Once again, it is more important that this planning branch forces decision makers to ask the right questions than to guarantee the right answers. For there are no "right" answers to the ultimate questions of environmental policy—only questions that breed more questions.

We do not mean to be too harsh in our judgment of the congressional draftsmen of 1970 or 1977. Even today, the hard work required to design a workable form of ends-oriented agency forcing is yet to be done. Encouraging signs can be seen, however, in Congress,[15] in the executive branch,[16] and in academia.[17] Given its experience with New Deal agencies, Congress was sensible to try to force aggressive agency action with the crude tools at its disposal. Like our predecessors, however, we too should try to learn from our mistakes and hope that our future blunders will be as fruitful as theirs were.

NOTES

CHAPTER 1

1. The total annual cost of air and water pollution abatement in 1978 alone was $39.5 billion. 10 U.S. COUNCIL ON ENVIRONMENTAL QUALITY, ANN. REP.: ENVIRONMENTAL QUALITY—1979 667 (1979). The CEQ estimates that air and water pollution control will cost $588 billion between 1970 and 1987. *Id.*

2. Pub. L. No. 90–604, 84 Stat. 1676 (current version at 42 U.S.C. §§ 7401–7642 (Supp. II 1978)) (formally entitled Clean Air Amendments of 1970).

3. 2 U.S. ENERGY INFORMATION AD., ANN. REP. TO CONGRESS 135 (1979) (1979 figures).

4. A recent study predicts that coal will be the primary energy source for 55% to 57% of the electricity produced in the United States by the year 2000. COAL—BRIDGE TO THE FUTURE: REPORT OF THE WORLD COAL STUDY 245 (C. Wilson ed. 1980). Due to increased overall demand for electric power by the turn of the century, this increased market share may well generate an increase of 100% to 200% in the amount of coal burned to generate electricity. *Id.*

5. *See* COAL—BRIDGE TO THE FUTURE: REPORT OF THE WORLD COAL STUDY, *supra* note 4, at 61–83 (global prospects for alternative energy sources); *cf.* ENERGY FUTURE: REPORT OF THE ENERGY PROJECT AT THE HARVARD BUSINESS SCHOOL (R. Stobaugh and D. Yergin, eds. 1979) (domestic prospects). The Stobaugh and Yergin report advocates increased use of alternative energy sources, such as solar energy. *Id.* at 216–33 (favoring "balanced" energy program). Nevertheless, it predicts a large and increasing role for coal. *Id.* at 232–33. For an intelligent critique of the Stobaugh and Yergin study, which gives even greater emphasis to coal, *see* Joskow, *America's Many Energy Futures—a review of* "Energy Future"; "Energy: The Next Twenty Years"; *and* "Energy in America's Future," 11 BELL J. ECON. 377, 383–84 (1980).

6. In 1976, power plants produced 65% of the sulfur oxides, 29% of the nitrogen oxides, and 25% of the particulates emitted in the United States. EPA, 1976 NATIONAL EMISSIONS REPORT vii (EPA Pub. No. 450/4–79–019, 1979). A longer-range threat is posed by the "greenhouse effect" that may be caused by the carbon dioxide generated

through massive increases in coal burning throughout the world. *See* W. NORDHAUS, THE EFFICIENT USE OF ENERGY RESOURCES 128–54 (1979).

7. 44 Fed. Reg. 33, 580–624 (1979) (codified in 40 C.F.R. § 60.43a (1979)). The controversy surrounding the 1979 NSPS has already generated a modest literature. *See, e.g.,* ENERGY: THE NEXT TWENTY YEARS 378–81 (Ford Foundation, 1979); Navarro, *The Politics of Air Pollution,* 59 PUBLIC INTEREST 39 (Spring 1980); Badger, *New Source Standard for Power Plants I: Consider the Costs,* 3 HARV. ENVT'L L. REV. 48 (1980); Ayres & Doniger, *New Source Standard for Power Plants II: Consider the Law,* 3 HARV. ENVT'L L. REV. 63 (1980). Although we agree with many points made by Navarro, Badger, and the Ford group, none of these studies attempts a full analysis of the decision's merit or an institutional analysis of the decision's decade-long development.

8. Sierra Club v. Costle, No. 79-1565 (D.C. Cir., filed June 11, 1979).

9. Although these principles have deep roots in American history, *see* M. BERNSTEIN, REGULATING BUSINESS BY INDEPENDENT COMMISSION 26–30, 37–39, 50–52 (1955), we follow the most thoughtful recent scholarship in viewing the New Deal as the period in which these ideals gained most practical importance and intellectual support. *See* J. FREEDMAN, CRISIS AND LEGITIMACY 32–33, 44–46, 59–60 (1978) (conception of New Deal ideal broadly consistent with model described in this book); Stewart, *The Reformation of American Administrative Law,* 88 HARV. L. REV. 1667, 1676–81 (1975) (same); *cf.* J. LANDIS, THE ADMINISTRATIVE PROCESS 23–24, 68–70 (1938) (contemporary account of New Deal ideals).

10. J. LANDIS, *supra* note 9, at 69–70.

11. We use the term *independent agency* to include those agencies whose administrators the president cannot remove without cause. *See* K. DAVIS, ADMINISTRATIVE LAW AND GOVERNMENT 16 (2d ed. 1975). Prominent independent agencies include the Interstate Commerce Commission (founded in 1887), the Federal Trade Commission (1914), the Federal Communications Commission (1934), the Securities and Exchange Commission (1934), and the Civil Aeronautics Board (1938). *Id.*

12. *See id.* at 17.

13. Although the states have often been celebrated as laboratories of innovation and experiment, there is a pervasive ambiguity concerning the kinds of local conditions to which they are expected to respond. On the one hand, the states may be viewed as *political bodies* that ought to be responsive to local values and constituencies slighted in national politics. On the other hand, the states may be viewed as convenient reposi-

tories for *expert bureaucracies* with superior sensitivity to local contexts. Although both these views support insulation from the vagaries of national politics, they diverge in their affirmative recommendations. For a sensitive treatment of the tension, *see* Stewart, *Pyramids of Sacrifice? Problems of Federalism in Mandating State Implementation of National Policy*, 86 YALE L.J. 1196, 1211–22 (1977); for an excellent legally oriented study of "cooperative" structures, *see* Mashaw, *The Legal Structure of Frustration: Alternative Strategies for Public Choice Concerning Federally Aided Highway Construction*, 122 U. OF PA. L. REV. 1 (1973); for critical views from other disciplines *see* A. WILDAVSKY & J. PRESSMAN, IMPLEMENTATION 87–124 (1973); Rose-Ackerman, *Risktaking and Reelection: Does Federalism Promote Innovation?* 9 J. OF LEGAL STUDIES 593–616 (1980).

14. *See, e.g.*, FCC v. Pottsville Broadcasting Co., 309 U.S. 134, 138–44 (1940) (Frankfurter, J.); J. FREEDMAN, *supra* note 9, at 45–46.

15. 2 K. DAVIS, ADMINISTRATIVE LAW TREATISE § 16.05 (1st ed. 1958) (requirement that agency state findings and reasons); 4 *id.* §§ 29.02, 29.03 ("substantial evidence" review on record as a whole).

16. *See* S. BREYER & R. STEWART, ADMINISTRATIVE LAW AND REGULATORY POLICY 288–90 (1979). A single paragraph cannot, of course, do justice to the complexity of the thought of scholars and judges who gave substance to this conception of administrative law, or to the countervailing tendencies that hid beneath its surface. Reading a few hundred pages, selected at random, from one of the great treatises of the last generation is the best way to gain a sense of the texture of the discourse. *See* K. DAVIS, *supra* note 15; L. JAFFE, JUDICIAL CONTROL OF ADMINISTRATIVE ACTION (1965).

17. Verkuil, *The Emerging Concept of Administrative Procedure*, 78 COLUM. L. REV. 258, 281–84 (1978).

18. Once again we shall not try to detail the trends in all their complexity. For a more complete survey of the regulatory reform movement, *see* ABA COMM'N ON LAW AND THE ECONOMY, FEDERAL REGULATION: ROADS TO REFORM 19–32 (1978 exposure draft).

19. *See, e.g.*, PRESIDENT'S ADVISORY COUNCIL ON EXECUTIVE ORGANIZATION: A NEW REGULATORY FRAMEWORK (1971) (the "Ash Council" report); L. KOHLMEIER, THE REGULATORS 73 (1969); and M. BERNSTEIN, *supra* note 9, at 180–85.

20. *See, e.g.*, J. SAX, DEFENDING THE ENVIRONMENT 52–56, 60–62, 107 (1970); T. LOWI, THE END OF LIBERALISM 72–93 (1969); M. BERNSTEIN, *supra* note 9, at 155–60.

21. PRESIDENT'S ADVISORY COUNCIL ON EXECUTIVE ORGANIZATION,

supra note 19, at 34–35; H. FRIENDLY, THE FEDERAL ADMINISTRATIVE
AGENCIES 142–47 (1960).

22. *See, e.g.*, H. FRIENDLY, *supra* note 21, at 13–14, 163–73 (need for
increased specificity in legislative definition of agency goals); Sive,
Some Thoughts of an Environmental Lawyer in the Wilderness of Administrative Law, 70 COLUM. L. REV. 612, 614–19 (1970) (advocating increased
judicial scrutiny of administrative actions).

23. *See, e.g.*, United States v. Florida East Coast Ry. Co., 410 U.S.
224, 240–42 (1973) (ratification of less formal agency procedures);
Scenic Hudson Preservation Conf. v. Federal Power Comm'n, 354 F.2d
608, 620–25 (2d Cir. 1965) (remand to conduct more detailed proceedings). *See generally* 1 K. DAVIS, ADMINISTRATIVE LAW TREATISE § 6.1 (2d
ed. 1978) (applauding widespread judicial innovation during 1970s in
review of agency rule making).

Congressional approaches have varied from attempts to influence
regulation by supplementing the information considered by regulators,
see, e.g., National Environmental Policy Act of 1969, Pub. L. No.
91–190, 83 Stat. 852 (1969) (codified at 42 U.S.C. §§ 4321–4347
(1976)), to the direct specification of technical standards, *see, e.g.*, 15
U.S.C. §§ 2001–2012 (1976) (automotive fuel efficiency standards), to
phased deregulation, *see, e.g.*, Airline Deregulation Act of 1978, Pub. L.
No. 96–504, 92 Stat. 1705 (1978) (to be codified at 49 U.S.C. §
1301–1308, 1324, 1371–1389). *See generally* SENATE COMM. ON GOVERNMENTAL OPERATIONS, STUDY ON FEDERAL REGULATION, SEN. DOC. NO.
25, 95th Cong., 1st Sess. (1977).

24. *See* ABA COMM'N ON LAW AND THE ECONOMY, *supra* note 18, at
27; S. BREYER & R. STEWART, *supra* note 16, at 283–85.

25. *See, e.g.*, J. ESPOSITO, VANISHING AIR 112–14 (1970); Stevens, *Air
Pollution and the Federal System: Responses to Felt Necessities*, 22 HASTINGS
L.J. 661, 679–81 (1971). For an opposing view with which we are in general agreement, *see* Jaffe, *The Administrative Agency and Environmental
Control*, 20 BUFFALO L. REV. 231, 233–36 (1970). Prof. Jaffe's skeptical
response to fashionable criticisms of the independent agency has a long
history. *See, e.g.*, Jaffe's review of M. BERNSTEIN, *supra* note 9, entitled
The Independent Agency: A New Scapegoat, 65 YALE L.J. 1068 (1956).

26. *See* O'Fallon, *Deficiencies in the Air Quality Act of 1967*, 33 LAW &
CONTEMP. PROB. 275, 286–96 (1968) (reviewing state agency resources
and performance); Vaughn, *State Air Pollution Control Boards: The Interest Group Model and the Lawyer's Role*, 24 OKLA. L. REV. 25, 30–32 (1971)
(inadequate staffing and insufficient budgets); J. ESPOSITO, *supra* note
25, at 125 (describing state agency as having "acted like an under-

ground chamber of commerce"). *See generally* Vaughn, *supra*, at 32–38, discussing methods of selecting state board members as well as agencies.

27. Pub. L. No. 91–604, 84 Stat. 1676 (codified at 42 U.S.C. §§ 1857b–1857l (1976)).

28. For a historical analysis of the passage of the act that emphasizes precisely the challenge to New Deal ideals, *see* Marcus, *Environmental Protection Agency*, in THE POLITICS OF REGULATION 267, 267–74 (J. Wilson ed. 1980) (citing relevant sources). For similar comments, *see* Stewart, *The Development of Administrative and Quasi-Constitutional Law in Judicial Review of Environmental Decisionmaking: Lessons from the Clean Air Act*, 62 IOWA L. REV. 713, 722–25 (1977). For an especially insightful analysis by a political scientist *see* Ingram, *The Political Rationality of Innovation: The Clean Air Act Amendments of 1970*, in A. FRIEDLAENDER, APPROACHES TO CONTROLLING AIR POLLUTION 12–53 (1978).

29. Until 1970, air pollution control was the responsibility of the National Air Pollution Control Administration (NAPCA), an administrative subdivision of the Department of Health, Education and Welfare. In July of that year, President Nixon submitted Reorganization Plan No. 3 of 1970 to Congress. Upon becoming effective, the plan created the Environmental Protection Agency (EPA) and transferred most of the functions formerly vested in NAPCA to the new agency. Reorg. Plan No. 3 of 1970, 3 C.F.R. 1072 (1966–1970 Compilation) *reprinted in* 5 U.S.C. app. at 827 (1976) *and in* 84 Stat. 2086 (1970).

30. Clean Air Amendments of 1970, § 304, 42 U.S.C. § 7604 (Supp. II 1978).

31. Clean Air Amendments of 1970, § 109, amended and codified at 42 U.S.C. § 7409(b)(1) (1976). The act refers to health-related standards as "primary" ambient air quality standards. It also calls for the establishment of "secondary" ambient air quality standards, which are, in the administrator's judgment, requisite to protect the public "welfare." These welfare-related standards are intended to prevent harm to plant and animal life, as well as to soil and water and aesthetic values such as visibility. *See* 42 U.S.C. § 7602(h) (Supp. II 1978). Air quality standards designed to protect the public health and welfare have so far been established for seven pollutants: sulfur oxides (SO_x); particulate matter; carbon monoxide (CO); ozone (O_3); hydrocarbons (HC); nitrogen dioxide (NO_2); and lead (Pb). 40 C.F.R. § 50.4–.12 (1979).

Unlike primary standards, *see* note 32 *infra*, the attainment of secondary standards carries no statutory deadline. Instead, the act requires attainment of secondary standards only within a "reasonable time." 42

U.S.C. § 7410(a)(2)(A) (Supp. II 1978). As a consequence of the failure
to specify enforceable deadlines, secondary standards have not received
great attention in the act's implementation.

32. For stationary sources such as power plants, the act originally
specified attainment of its health-related targets by 1975. *See* 42 U.S.C.
§ 1857c–5(a)(2)(A) (1976). Heavily conditioned provisions could, however, stretch these formal deadlines by an additional two years, *see*
Clean Air Act, § 110, 42 U.S.C. § 7410(e) (Supp. II 1978), and individual plants might qualify for an extra year's grace. 42 U.S.C. § 1857c–
5(f) (1976) (amended 1977).

During the early years of the act, the formal extensions provided under § 110 may have been of less practical importance (for individual
sources) than the informal, administrative extensions EPA provided
under § 113 of the act, 42 U.S.C. § 1857c–8 (1976) (amended 1977 and
recodified at 42 U.S.C. § 7413). *See* Currie, *Relaxation of Implementation
Plans under the 1977 Clean Air Amendments*, 78 MICH. L. REV. 155, 157–59
(1979); W. Rodgers, ENVIRONMENTAL LAW 345–46 (1977). The legality
of the use of § 113 to evade the act's timetables was never firmly established, *see id.* at 347, and elicited criticism both from commentators, *see
id.* at 346–47, and the Congress, *see* S. REP. NO. 127, 95th Cong., 1st
Sess. 45 (1977). The Clean Air Act Amendments of 1977 largely eliminated this potential loophole.

33. 42 U.S.C. § 7410(a) (Supp. II 1978). *See also* W. RODGERS, ENVI-
RONMENTAL LAW 233–35 (1977) (discussion of state processes and available federal assistance). The critical feature of the SIP process has been
its reliance on air quality models that predict how a plant's emissions affect the air quality in its surrounding area. Due to these models' technical complexity, EPA has provided states with a massive set of support
documents as aids in the preparation of SIPs. *See id.* at 235–37.

34. 42 U.S.C. § 7410(a)(2) (Supp. II 1978). *See* Natural Resources
Defense Council v. Train, 421 U.S. 60, 93–94 (1975). For areas already
in compliance with the ambient air standards, the doctrine of prevention of significant deterioration (PSD) came to require, in addition, the
maintenance of existing levels of air quality. *See* Sierra Club v.
Ruckelshaus, 344 F. Supp. 254 (D.D.C.), *aff'd*, 4 Envir. Rep. Cas. 1815
(D.C. Cir. 1972), *aff'd by an equally divided court sub nom.* Fri v. Sierra
Club, 412 U.S. 541 (1973). For further discussion of PSD and its impact
on NSPS, *see* pp. 28–33 *infra*.

35. Statewide average SIP requirements for coal-burning power
plants range from 0.8 to 5.0 pounds of SO_2 per million BTU produced.
Memorandum from Tommy Holland, Energy Information Section, to

John H. Haines, Assistant to Director, Emissions Standards and Engineering Division, EPA, 3 (May 23, 1978) (EPA Docket No. OAQPS-78-I, Item No. II-B-117). Individual plants in some states are allowed emissions ranging from 8 to 10 pounds per MBTU (using short-term averages). *See, e.g.*, 40 C.F.R. §§ 52.1881(b)(13)(i), (b)(23), (b)(28), (b)(33) (1979) (emission limits, set by EPA directly, for Ohio counties and sources).

36. Many states initially failed, or refused, to take advantage of the act's provisions which allowed them to tailor SIPs to individual plants. Instead, states typically required uniform statewide rollbacks in emissions. *See* Memorandum from Walter C. Barber, Director, Office of Air Quality Planning and Standards, EPA, to Barbara Blum, Deputy Administrator, EPA (Dec. 6, 1979), *reprinted in* 10 ENVIR. REP. (BNA) 1873 (1980). As a result, individual plants were often subject to requirements more stringent than necessary to meet federal clean-up targets. In response, EPA urged many states to develop more flexible, plant-by-plant requirements. For a sobering assessment of the SIP process, *see* Roberts & Farrell, *The Political Economy of Implementation: The Clean Air Act and Stationary Sources*, in A. FRIEDLAENDER, APPROACHES TO CONTROLLING AIR POLLUTION 152 (1978).

37. 42 U.S.C. § 7411 (Supp. II 1978) (amending 42 U.S.C. § 1357c–6 (1976)).

38. Clean Air Amendments of 1970 § 111(a)(1), 42 U.S.C. § 1857c–6(a)(1) (1976).

39. For example, control technology necessary to remove SO_2 from smokestacks requires a relatively fixed amount of space within a plant near the smokestack. Because many plants built before 1970 did not anticipate the need to allow room for add-on technology, emission control is often more difficult and expensive for old plants than for new ones. Retrofitting typically increases the original cost of control technology by 25% to 30%. *See* NATIONAL ACADEMY OF SCIENCES, AIR QUALITY AND STATIONARY SOURCE EMISSION CONTROL, PREPARED FOR HOUSE COMM. ON PUBLIC WORKS, 94th Cong., 1st Sess. 451–54 (Comm. Print 1975) [hereinafter cited as STATIONARY SOURCE EMISSION CONTROL].

CHAPTER 2

1. Clean Air Amendments of 1970, § 111(a)(1), 42 U.S.C. § 1857c–6(a)(1) (1976).

2. Courts have, with varying force, supported this suggestion. Under a provision of the Federal Water Pollution Control Act, 33 U.S.C. §

1311(b)(2)(A) (1976) (limitations that require "best available technology economically achievable for such category or class, which will result in reasonable further progress toward the national goal of eliminating the discharge of all pollutants. . . ."), closely analogous to Section 111 of the Clean Air Act, the Fourth Circuit has held that EPA must compare the costs and ecological benefits of its chosen standard with those of alternative levels of heat reduction, or, at the very least, provide the "best information available" on the expected ecological benefits of the ordered reduction. Appalachian Power Co. v. Train, 545 F.2d 1351, 1364 (4th Cir. 1976); *accord,* National Crushed Stone Ass'n v. EPA, 601 F.2d 111, 121 (4th Cir. 1979), *cert. granted,* 100 Sup. Ct. 1011 (1980).

Although the Court of Appeals for the District of Columbia has not required EPA to conduct cost-benefit analyses as part of the establishment of NSPS, Portland Cement Ass'n v. Ruckelshaus, 486 F.2d 375, 387 (D.C. Cir. 1973), *cert. denied,* 417 U.S. 921 (1974), it has required the agency to establish a minimal relationship between costs and expected benefits, Essex Chemical Corp. v. Ruckelshaus, 486 F.2d 427, 433 (D.C. Cir. 1973), *cert. denied,* 416 U.S. 969 (1974) ("best system" in Section 111 means system that "can reasonably be expected to serve the interests of pollution control without becoming exorbitantly costly in an economic or environmental way"). The D.C. Circuit has also required that EPA consider cost-benefit analyses submitted to it. Portland Cement Ass'n v. Ruckelshaus, 486 F.2d at 387.

3. For assessments of coal cleaning processes written during the early 1970s, *see* STATIONARY SOURCE EMISSION CONTROL, *supra* note 39, chap. 1, at 370–72; MITRE CORP., THE PHYSICAL DESULFURIZATION OF COAL 41 (1970) (Nat'l Tech. Inf. Serv. No. PB–210–373). Coal cleaning provides a number of benefits other than the removal of sulfur from coal. *See* note 32, chap. 5 *infra.* As early as 1960, well before the widespread adoption of pollution control regulations, more than half of U.S. coal production was cleaned to some degree. *Id.* at 41. An excellent contemporary overview of coal cleaning processes appears in a paper by J. Kilgroe, Coal Cleaning for Sulfur Oxide Emission Control (Apr. 8–9, 1980) (unpublished paper presented at Acid Rain Conference, Springfield, Va.). More technical descriptions of cleaning can be found in Huettenhain, Yu, & Wong, *A Technical and Economic Overview of Coal Cleaning,* in I PROCEEDINGS: SYMPOSIUM ON COAL CLEANING TO ACHIEVE ENERGY AND ENVIRONMENTAL GOALS 256, 259–74, 277–85 (1979) (EPA Pub. No. 600/7–79–098a) [hereinafter cited as SYMPOSIUM ON COAL CLEANING]. A detailed discussion of the costs of conventional coal cleaning techniques appears at pp. 67–69 *infra.*

More advanced physical coal cleaning processes now being developed grind coal into a finer slurry and use more sophisticated separation techniques. In addition, methods are being developed to reach and remove sulfur that is chemically bonded within the coal. *See* J. Kilgroe & J. Strauss, Use of Coal Cleaning for Air Quality Management, 7–15 (Jan. 22–25, 1980) (unpublished paper presented at Second Conference on Air Quality Management in the Electric Power Industry). Although many of these processes are technologically promising, it is not yet known whether they will be economically competitive with other control technologies such as the scrubber. In addition, all coal cleaning processes, but especially chemical coal cleaning processes, produce waste products that may cause second-order environmental problems unless properly treated. *Id.*

4. The first full-scale scrubber built in the United States began operation in 1968 and was abandoned in 1971. By the end of 1971, the nation's second and third scrubbers had begun operation. EPA, REPORT OF THE HEARING PANEL: NATIONAL PUBLIC HEARINGS ON POWER PLANT COMPLIANCE WITH SULFUR OXIDE AIR POLLUTION REGULATIONS 88–89 (1974) [hereinafter cited as REPORT OF THE HEARING PANEL].

5. *See* Torstrick, Henson, & Tomlinson, *Economic Evaluation Techniques, Results, and Computer Modeling for Flue Gas Desulfurization*, in I PROCEEDINGS: SYMPOSIUM ON FLUE GAS DESULFURIZATION 118, 121, 134 (1978) (EPA Pub. No. 600/7–78–058a) [hereinafter cited as FGD SYMPOSIUM]. The brief technical discussion of scrubbing that follows in the text is largely derived from II EPA, FLUE GAS DESULFURIZATION SYSTEM CAPABILITIES FOR COAL-FIRED STEAM GENERATORS 3–2 to 3–111, 4–1 to 4–34 (1978) (EPA Pub. No. 600/7–78–032b) [hereinafter cited as FGD SYSTEM CAPABILITIES].

6. FGD SYSTEM CAPABILITIES, *supra* note 5, at 4–13 (approximately 30% operability for systems built before 1973). Despite significant improvements since the early 1970s, *id.*, breakdowns still trouble many plants. *See* EPA, EPA UTILITY FGD SURVEY: DECEMBER 1978–JANUARY 1979 x (1979) (EPA Pub. No. 600/7–79–022c) (in survey of plant operation over two-month period, half of systems had less than 50% reliability during at least one month).

7. *See* 37 Fed. Reg. 5768–69 (1972); REPORT OF THE HEARING PANEL, *supra* note 4, at 88.

8. *See* EPA, BACKGROUND INFORMATION FOR PROPOSED NEW SOURCE PERFORMANCE STANDARDS 10–16 (1971) (coal-fired power plants) [hereinafter cited as 1971 BACKGROUND DOCUMENT] (including statement that proposed standards are "insensitive to cost/benefit analysis when

stack-gas cleaning is employed, in that it is the only control system available").

9. 37 Fed. Reg. 5768–69 (1972).

10. The British thermal unit (BTU) is a measure of the amount of energy released by coal when burned. Typical bituminous coal contains roughly 22 million BTU (MBTU) per ton.

The amount of sulfur found in coal may be expressed either as a percentage of the total weight of the coal or as the amount of sulfur dioxide that will be released per MBTU of heat generated by the coal. Because coal varies substantially in heat content per pound, coals with the same percentage of sulfur by weight may yield different amounts of SO_2 while producing the same amount of energy. The measurement of sulfur content in terms of pounds of sulfur dioxide produced per MBTU is more useful for determining how "dirty" or "clean" a coal is to burn and will be used here whenever possible.

11. *See* McCreery & Goodman, *An Evaluation of the Desulfurization Potential of U.S. Coals*, in I SYMPOSIUM ON COAL CLEANING, *supra* note 3, at 387, 404 (characterizing coals on low end of scale); EPA, SULFUR REDUCTION POTENTIAL OF U.S. COALS 28 (1976) (EPA Pub. No. 600/2–76–091) (fig. 16) (reprinting U.S. Dep't of Interior Bureau of Mines, Pub. No. RI 8118) (showing extent of U.S. reserves containing more than 10 pounds of sulfur per MBTU).

American coal reserves are enormous, and there is little work that provides detailed characterizations of the combined sulfur content, heat content, and sulfur reduction potential of the United States reserve base. *See* EPA, ELECTRIC UTILITY STEAM GENERATING UNITS: BACKGROUND INFORMATION FOR PROPOSED SO_2 EMISSION STANDARDS 4–3 (1978) (EPA Pub. No. 450/2–78–007a) [hereinafter cited as 1978 SO_2 BACKGROUND INFORMATION]. Although the U.S. Bureau of Mines has in its records a large number of detailed samples of coal reserves from various locations, further analysis is required to determine how representative these samples are of the entire reserve base. *See* McCreery & Goodman, *supra*, at 403–06 (reporting high degree of uncertainty in representativeness of previous studies).

12. U.S. ENERGY INFORMATION ADMINISTRATION, *supra* note 3, chap. 1, at 123 (236 billion of 438 billion tons of coal in national demonstrated reserve base lie west of Mississippi River; figures based on extractable amount of heat unavailable, might show lower proportion of national reserves because of lower heat content of many western coals). Roughly 64% to 70% of all western coal produces less than 1.2 pounds of SO_2 per MBTU. *See* 1978 SO_2 BACKGROUND INFORMATION, *supra* note 11, at

4–6. Western coal has accounted for a small, but increasing, share of national production. 2 U.S. Dep't of Interior, Bureau of Mines, Minerals Yearbook 1961 59 (1962); National Coal Association, Coal Facts 80–81 (1978–79) (5% in 1961; 9% in 1971; 24% in 1977) (all figures exclude Alaska).

13. Only 7% to 10% of the 109 billion ton recoverable reserve in the Eastern and eastern Midwest regions produces less than 1.2 pounds of SO₂ per MBTU. 1978 SO₂ Background Information, *supra* note 11, at 4–6.

14. *See* 42 U.S.C. § 1857c–6(b) (duty of administrator to establish "standards of *performance*" [emphasis added]). *Cf.* S. Rep. No. 1196, 91st Cong., 2d Sess. 17 (1970) (administrator "should not make a technical judgment as to how the standard [of performance] should be implemented. He should determine the achievable limits and let the owner or operator determine the most economic, acceptable technique to apply.")

15. *See generally, e.g.*, A. Kneese, R. Ayres, & R. D'Arge, Economics and the Environment 84–85 (1970).

16. *See* 1971 Background Document, *supra* note 8, at 15–16.

17. 37 Fed. Reg. 5767–71 (1972) (supplemental statement in connection with final promulgation; published after Kennecott Copper Corp. v. EPA, 462 F.2d 846 (D.C. Cir. 1972) (requiring more specific explanations of reasoning underlying EPA ambient air regulations)).

18. 1971 Background Document, *supra* note 8, at 12–13 (low-sulfur coal could be burned to meet standard, but not enough of a supply for all new and existing plants, and too expensive to ship to some areas).

19. *See id.* at 16.

20. *Id.* at 12.

21. Essex Chemical Corp. v. Ruckelshaus, 486 F.2d 427, 441 (D.C. Cir. 1973), *cert. denied*, 416 U.S. 969 (1974). A single lime-limestone scrubbing system may produce several hundred thousand tons of sludge a year. EPA, Controlling SO₂ Emissions from Coal-Fired Steam-Electric Generators: Solid Waste Impact 14–18 (1978) (EPA Pub. No. 600/7–78–044a).

The court's footnote mention of certain antipollution possibilities, *see* 486 F.2d at 440 n.48 (use of low-sulfur coal complies with emission standard without scrubbing); *id.* at 441 n.51 (speculating on desirability of more flexible standard), indicated a broader awareness of the agency's failings—but the court did not act upon this perception in a legally effective fashion.

22. 40 Fed. Reg. 42045, 42047 (1975) (draft response to remand); 42

Fed. Reg. 61541 (1977) (final promulgation of standards, declaring that "none of the comments persuaded the Agency to change the Policy expressed in the September, 1975, draft response, i.e., that scrubber sludge can be fixated and disposal [*sic*] of in an environmentally acceptable manner at reasonable costs.")

23. Oljato Chapter of the Navaho Tribe v. Train, 515 F.2d 654 (D.C. Cir. 1975).

24. *See* Letter from Joseph J. Brecher, Native American Rights Fund, to William D. Ruckelshaus, Administrator, EPA 3 (March 20, 1973) (request on behalf of Indian tribes that EPA issue notice of proposed rule making to revise 1971 NSPS for power plants).

25. In addition to arguing that scrubbers be required, the Navaho questioned EPA's finding that scrubbers could remove approximately 70% of SO_2, insisting that scrubbers were capable of 90% sulfur removal. We use EPA's determination of demonstrated capabilities in the present example.

26. *See* pp. 70–72 *infra*.

27. *See, e.g.*, Letter from John R. Quarles, Jr., Acting Administrator, EPA, to Martin Green, Dep't of Justice (Jan. 23, 1974).

28. Oljato Chapter of the Navaho Tribe v. Train, 515 F.2d 654, 661, 666 (D.C. Cir. 1975). The Navaho attempted to sue the agency under § 304 of the act, 42 U.S.C. § 7604 (Supp. II 1978), on the theory that failure to revise an arguably invalid standard constituted a failure to fulfill a nondiscretionary duty. The court rejected this argument but indicated that the Navaho might obtain judicial review of the NSPS after petitioning the administrator under § 307. Following this initial defeat, the Navaho, joined by the Sierra Club, renewed their efforts to reduce western emissions by following a different procedural course. In early 1977, EPA agreed to the groups' request for formal reconsideration of the 1971 NSPS. *See* 42 Fed. Reg. 5121 (1977). Reconsideration had barely begun, however, when Congress enacted its own revisions to § 111 as part of the Clean Air Act Amendments of 1977, Pub. L. No. 95–95, 91 Stat. 685 (codified at 42 U.S.C. § 74 (Supp. II 1978)). *See* chapters 3 and 4 *infra*.

29. EPA, 1973 NATIONAL EMISSIONS REPORT 1 (1974) (EPA Pub. No. 450/2–74–012).

30. The compliance problems of existing power plants were widely studied at this time. *See, e.g.*, REPORT OF THE HEARING PANEL, *supra* note 4; ADMINISTRATOR OF EPA, CONTROL OF SULFUR OXIDES, S. DOC. NO. 59, 94th Cong., 1st Sess. (1975) [hereinafter cited as CONTROL OF SULFUR OXIDES]; EPA, POSITION PAPER ON REGULATION OF ATMOSPHERIC

SULFATES (1975) (EPA Pub. No. 450/2–75–007) [hereinafter cited as POSITION PAPER]; STATIONARY SOURCE EMISSION CONTROL, *supra* note 39, chap. 1, at 193–711 (1975). Only EPA's *Position Paper* gave more than a glance at the problems posed by strategies to control new plants. *See* POSITION PAPER, *supra*, at xvi–xix, 60–80.

31. In the fall of 1973, 70% of the nation's coal and oil-burning power plants emitted more SO_2 than allowed by the proposed SIPs. EPA, EPA ENFORCEMENT: A PROGRESS REPORT: 1976 12 (1977). The cutbacks needed were so large that it was already apparent that most coal-burning power plants would not meet the original mid-1975 deadline. REPORT OF THE HEARING PANEL, *supra* note 4, at 3. By 1975, the EPA estimated that at least 40% of the coal-burning capacity in the U.S. still required emission reductions to comply with primary air quality standards. REPORT OF THE HEARING PANEL, *supra* note 4, at 14. The EPA was no longer planning for compliance by June of that year or even by 1977, but was looking to 1980. CONTROL OF SULFUR OXIDES, *supra* note 30, at 1–6.

32. 42 U.S.C. § 1857c–5(a)(2) (1976). *See* W. RODGERS, *supra* note 32, chap. 1, at 235–37 (description of complex factors on which SIP emission limitations are based).

33. *See* STATIONARY SOURCE EMISSION CONTROL, *supra* note 39, chap. 1, 212–16. In 1975, a tall stack was estimated to cost 3% to 17% the price of a scrubber. *Id*. at 210, 215. At least fifteen states provided credit for the use of tall stacks in 1974. *See Clean Air Act Oversight: Hearings before the Subcomm. on Environmental Pollution of the Senate Comm. on Public Works*, 93d Cong., 2d Sess. 320, 330 (1974) (statement of Richard Ayres) [hereinafter cited as *1974 Senate Oversight Hearings*].

34. *See, e.g., Implementation of the Clean Air Act—1975: Hearings before the Subcomm. on Environmental Pollution of the Senate Comm. on Public Works, Pt. 2*, 94th Cong., 1st Sess. 1417–21 (1975) (statement of Donald G. Allen, New England Electric System, presenting discussion by James R. Mahoney, consultant) [hereinafter cited as *1975 Implementation Hearings*].

35. *See, e.g.*, Ayres, *Enforcement of Air Pollution Controls on Stationary Sources under the Clean Air Amendments of 1970*, 4 ECOLOGY L. Q. 441, 452–54 (1975) (describing "environmental bankruptcy" of dispersion strategy); *Clean Air Act Amendments of 1977: Hearings before the Subcomm. on Health and the Environment of the House Comm. on Interstate and Foreign Commerce*, 95th Cong., 1st Sess. 493 (1977) (statement on behalf of Izaak Walton League of America) (opposing tall stacks because of acid rain and sulfates) [hereinafter cited as *1977 House Hearings*]; *1975 Im-*

plementation Hearings, supra note 34, at 1899 (letter from League for Conservation Legislation).

For an excellent discussion of the enforcement problems presented by ICS, *see* 120 Cong. Rec. 18960 (1974) (reprinting memorandum from EPA enforcement division). Environmental groups based their objections, in part, on the difficulty of proving that a utility had failed to burn low-sulfur coal or comply with a complex dispatching algorithm. *See 1974 Senate Oversight Hearings, supra* note 33, at 320, 325–26 (statement of Richard Ayres); *1975 Implementation Hearings, supra* note 34, at 1313–14 (statement of Benjamin Wakes, Montana Dept. of Health and Environmental Sciences).

36. *See, e.g.,* NRDC v. EPA, 489 F.2d 390, 406–09 (5th Cir. 1974) *rev'd in part on other grounds sub nom.* Train v. NRDC, 421 U.S. 60 (1975) (Clean Air Act permits use of dispersion techniques such as tall stacks only where methods of emission reduction unavailable or unfeasible).

37. Position Paper, *supra* note 30, at xv–xix.

38. For example, Ayres criticized the EPA's attempt to use complex diffusion models to produce a closer fit between ambient air criteria and control requirements. *See* Ayres, *supra* note 35, at 468–69.

39. *See, e.g., 1974 Senate Oversight Hearings, supra* note 33, at 320–24 (statement of Richard Ayres); *1975 Implementation Hearings, supra* note 34, at 1899 (letter from League for Conservation Legislation); Ayres, *supra* note 35, at 443–49.

40. By the end of 1976, only 59% of the nation's power plants were operating in full compliance with sulfur oxides emission limitations or meeting federally enforceable schedules. EPA, EPA Enforcement: A Progress Report, 1976, *supra* note 31, at 12. Of the plants comprising the remaining 41%, many operated in areas where challenges to SIPs remained unresolved. *Id.*

41. *1975 Implementation Hearings, supra* note 34, at 1421 (statement of Donald Allen, New England Electric System); *id.* at 1431, 1439–40 (statement of Electric Utility Industry Clean Air Coordinating Committee); *1977 House Hearings, supra* note 35, at 1365, 1383 (statement of A. Joseph Dowd on behalf of Edison Electric Institute); 6 Envir. Rep. (BNA) 130 (1975) (utility representatives' statements to panel of National Governors Conference energy officials); *see* note 39 *supra* (for views of environmentalists).

42. *See, e.g., Clean Air Act Oversight—1973: Hearing before the Subcomm. on Public Health and Environment of the House Comm. on Interstate and Foreign Commerce,* 93d Cong., 1st Sess. 13, 16–17 (1973) (testimony of John A. Quarles, Jr., Acting Administrator, EPA); *Clean Air Act Amend-*

ments—1975: Hearings before the Subcomm. on Health and the Environment of the House Comm. on Interstate and Foreign Commerce, 94th Cong., 1st Sess. 34, 39–40 (1975) (testimony of Russell E. Train); *id*. at 826–27 (testimony of representative of Nat'l Coal Association) [hereinafter cited as *House Hearings—1975 Amendments*].

43. *See, e.g.*, Sierra Club v. Train, No. 76–0656 (D.C. Cir., *dismissed* January 19, 1978) (suit to establish ambient air quality standard for sulfates); Ethyl Corp. v. EPA, 541 F.2d 1 (D.C. Cir. 1976), *cert. denied*, 426 U.S. 941 (1977) (participation in litigation establishing ambient air quality standard for lead).

CHAPTER 3

1. *Cf.* J. LANDIS, *supra* note 9, chap. 1, at 68–69 (detailed delegation requires repeated amendment).

2. *See* 42 U.S.C. § 1857f–1(b) (1976). *Cf.* 42 U.S.C. § 1857f–1(b) (1970) (original deadlines for mobile sources). Congress first extended deadlines for automobiles in 1974, as part of its initial reaction to the OPEC oil embargo of 1973. *See* Energy Supply and Environmental Coordination Act of 1974, Pub. L. No. 94–319, § 5, 88 Stat. 258.

3. *See* 42 U.S.C. § 1857c–5 (1976). Although the agency had temporarily found ways to provide individual polluters with additional time to meet ambient air quality standards, *see* Currie, *supra* note 32, chap. 1, at 155–59 (account of EPA use of delayed compliance orders under § 113 of the act), it was questionable whether, and for how long, the original, highly specific, statutory timetables would allow for continued delay. *See* W. RODGERS, *supra* note 32, chap. 1, 345–47. *But cf.* Union Electric Co. v. EPA, 427 U.S. 246, 267–68 (1976) (claims of technological or economic infeasibility relevant to fashioning compliance orders under § 113). In 1976, EPA announced that it would require final adjustments to state implementation plans by July 1, 1977. *See* 7 ENVIR REP. (BNA) 455 (1976).

4. *See, e.g., Oversight—Clean Air Act: Hearing before the Subcomm. on Public Health and Environment of the House Comm. on Interstate and Foreign Commerce*, 93d Cong., 2d Sess. 9–10 (1974) (statement of Russell E. Train, Administrator, EPA) [hereinafter cited as *1974 House Oversight Hearing*]; *House Hearings—1975 Amendments, supra* note 42, chap. 2, at 34, 39–40; *id*. at 826–27.

A related topic of interest in the mid-1970s concerned the control measures that were to be applied to power plants receiving federal orders to convert from oil to coal. Here environmentalists won a significant legislative victory. Under the Energy Supply and Environmental Coordination Act of 1974, the EPA could prevent coal conversion plants from increasing their emissions of sulfur oxides, regardless of whether resulting SO_2 concentrations violated local air quality standards. 42 U.S.C. § 1857c–10(d). A discussion of the dangers of sulfur oxides other than SO_2 appears in chap. 5, *infra*. For an excellent summary of the problems posed by coal-conversion legislation, see Ayres, *supra* note 35, chap. 2, at 447–49.

5. *See, e.g., House Hearings—1975 Amendments, supra* note 42, chap. 2, at 709–10 (statement of Richard Ayres).

6. The authors obtained the background information supplied here and below through interviews conducted in 1979–80 with Senate and House staffers. Individual citations are omitted in order to preserve confidentiality.

7. Sierra Club v. Ruckelshaus, 344 F. Supp. 253 (D.D.C.), *aff'd*, 4 Envir. Rep. Cas. 1815 (D.C. Cir. 1972), *aff'd by an equally divided court sub nom*. Fri v. Sierra Club, 412 U.S. 541 (1973).

8. *See, e.g., 1975 Implementation Hearings, supra* note 34, chap. 2, at 862–65 (statement of J. D. Geist, Exec. V.P., New Mexico Public Service Co.) (need to amend 1970 act to negate Sierra Club v. Ruckelshaus interpretation and avoid preclusion of "further development of my State's vast energy resources"); *id*. at 1984–85 (statement of E. Allan Hunter, President, Utah Power and Light Co.) (Sierra Club v. Ruckelshaus means "continued poverty in many rural areas of the west" and "inability to develop our own resources").

9. Clean Air Act Amendments of 1970, § 111(a)(1), 42 U.S.C. § 1853–6(a)(1) (prior to 1977 amendment) (emphasis added).

10. H.R. 10498, § 111(a)(1)(C), *reprinted in* H.R. REP. No. 1175, 94th Cong., 2d Sess. 310 (1976) (enacted as Pub. L. No. 95–95, sec. 109, § 111(a)(1)(C) (1977)) [hereinafter cited as 1976 HOUSE REPORT] (emphasis added).

11. *Id*. § 111(a)(7), *reprinted in 1976* HOUSE REPORT, *supra* note 10, at 310–11 (enacted as Pub. L. No. 95–95, sec. 109, § 111(a)(7) (1977)) (emphasis added). According to the *House Report*, this definition refers "to any technology which is inherently low polluting or nonpolluting (e.g., fluidized bed combustion, use of water-based paints instead of solvents, etc.)." *Id*. at 161.

12. The report states that a source "may no longer meet NSPS requirements merely by use of untreated oil or coal," because it is not "technological." 1976 HOUSE REPORT, *supra* note 10, at 161.

13. *See* Frankfurter, *Some Reflections on the Reading of Statutes*, 47

Colum. L. Rev. 526, 543 (1947).

14. Interview with representative of Electric Utility Industry Clean Air Coordinating Committee in New Haven, Conn. (Oct. 22, 1979).

15. *See, e.g., Nondegradation Policy of the Clean Air Act: Hearing before the Subcomm. on Air and Water Pollution of the Senate Comm. on Public Works*, 93d Cong., 1st. Sess. 125–36 [hereinafter cited as *Nondegradation Hearing*]; *House Hearings—1975 Amendments, supra* note 42, chap. 2, at 826 (statement of Robert V. Price, Nat'l Coal Association).

16. In 1980, 94% of the UMW's membership lived in states east of the Mississippi. (Information supplied by United Mine Workers of America.)

17. Interview with representative of National Coal Association.

18. 1976 House Report, *supra* note 10, at 160.

19. Many analysts have questioned whether § 111, even in its original form, promoted the development of new technologies. *See, e.g.*, Ayres, *supra* note 35, chap. 2, at 476 (best technology approach generates a "negative incentive" for the development of new controls); *cf.* Mills & White, *Government Policy toward Automotive Emissions Control* in A. Friedlaender, Approaches to Controlling Air Pollution 342 (1978) (same). And a percentage treatment requirement threatens to distort further the process of technological innovation. Such requirements *discourage* the development of technologies which fail to meet specified levels of percentage reduction. Preclusion of new processes may occur even though the application of such technologies to lower sulfur fuels might produce fewer emissions at lower cost than the use of approved technologies on the high-sulfur fuel that would otherwise be used. *See generally Oversight: Effect of the Clean Air Act Amendments on New Energy Technologies and Resources: Hearings before the Subcomm. on Fossil and Nuclear Energy Research, Development and Demonstration of the House Comm. on Science and Technology*, 95th Cong., 2d Sess. 126–30 (1978) (statement of James L. Liversman, Dept. of Energy) (percentage reduction requirement may inhibit development of emerging technologies).

20. *See* EPA, A Preliminary Analysis of the Economic Impact on the Electric Utility Industry of Alternative Approaches to Significant Deterioration (1976) (Nat. Tech. Information Serv. No. PB 251 394) [hereinafter cited as EPA Preliminary Analysis]; EPA & Federal Energy Admin., An Analysis of the Impact on the Electric Utility Industry of Alternative Approaches to Significant Deterioration (1975) (Nat. Tech. Information Ser. No. PB 246 205) [hereinafter cited as EPA–FEA Analysis].

21. *See* EPA Preliminary Analysis, *supra* note 20, at I–1 to I–4; EPA–FEA Analysis, *supra* note 20, at 1–3.

22. *See* EPA Preliminary Analysis, *supra* note 20, at V–17 (analysis of nationwide emissions only).

23. *Id.* at V–17. EPA predicted that a universal scrubbing requirement would result in a 1990 nationwide emissions level 22% lower than the level that would result from a 1.2 standard. Even without forced scrubbing, emissions were predicted to decline relative to 1974 levels.

In a separate analysis prepared for the administrative reconsideration of the 1971 NSPS that was requested by the Navaho and the Sierra Club, EPA analysts did compare the use of a lower ceiling to the use of a mandatory percent reduction. Strikingly, they concluded that "[i]f the decision is made to revise the NSPS . . . , then the most cost-effective and advantageous approach would be to retain the form of the existing standard and to lower the allowable emission limit," and that "the principal advantages of this approach [*i.e.* lowering the emission ceiling], other than cost considerations, are that it would reduce SO_2 emissions most in the high-sulfate region [the East] and that it would allow new plants latitude in determining how they will comply with the regulation." J. Crenshaw, D. Kirchgessner, H. Kuo, C. Miranda, & A. Wehe, Staff Study: Alternatives for Revising the SO_2 New Source Performance Standard for Coal-Fired Steam Generators at i, 33 (Oct. 22, 1976) (EPA Docket No. OAQPS–78–I, Item No. II–A–7). This more sophisticated analysis, however, was not part of the EPA's presentation to Congress.

24. EPA Preliminary Analysis, *supra* note 20, at V–8.

25. *Id.* at V–5 (figures for eastern production include, for these purposes only, coal produced in Arkansas and Oklahoma).

26. *Nondegradation Hearing, supra* note 15, at 141–49 (analysis by Nat'l Coal Association for markets in 100-mile strip of coastal area between Virginia and Maine).

27. *See* U.S. Dep't of Interior, Bureau of Mines, Minerals Yearbook 59 (1964); U.S. Dep't of Interior, Bureau of Mines, Minerals Yearbook 326 (1973).

28. *See, e.g., Clean Air Act Amendments 1977: Hearing before the Subcomm. on Environmental Pollution of the Senate Comm. on Environment and Public Works*, 95th Cong., 1st Sess. pt. 2, 192–216 (1977) (prepared statement of Richard E. Ayres) [hereinafter cited as *Senate Hearings—1977 Amendments*]. To the extent that the cited passages focus on NSPS, they almost invariably refer to the use of best available control technology in western, clean air areas. *See id.* at 208–10. *See also House Approves Air Bill Revising Panel's Rules on Autos, Clean-Air Areas*, 8 Envir.

Rep. (BNA) 155 (1977) (detailing legal attempts to weaken clean air provisions of 1977 amendments).

29. *See* Ayres, *supra* note 35, chap. 2, at 476 (agency development of technological expertise duplicates industry's activities, discourages innovation by industry, and distracts agency from environmental focus).

30. *EPA Public Hearing on Coal-Fired Steam Generators SO₂ Emissions* I–9 (May 25, 1977) (EPA Docket No. OAQPS–78–I, Item No. II–G–2). *See id*. at 47, 58–59 (May 26, 1977) (Item No. II–G–3) (statement of Nancy Bartlit, representing New Mexico Citizens for Clean Air and Water, Inc., and American Lung Association).

31. *Senate Hearings—1977 Amendments, supra* note 28, at 180, 209 (statement of Richard E. Ayres). Ayres continues to make explicit use of dirty coal rhetoric in the present controversy, now on appeal to the D.C. Circuit. *See* the revealing article by Ayres & Doniger, *New Source Standard for Power Plants, II: Consider the Law*, 3 Harv. Envt'l L. Rev. 63, 78 (1979).

32. *See, e.g.*, Ayres, *supra* note 35, chap. 2, at 476 (qualified support of pollution tax to internalize costs).

33. *See, e.g.*, Ayres & Doniger, *supra* note 31, at 72 (appropriate ceiling, based on 95% scrubbing, would be 0.17 pounds per MBTU).

34. *See* pp. 97–100 *infra*.

35. 1976 House Report, *supra* note 10, at 496–97.

36. *See* 7 Envir. Rep. (BNA) 532–34 (1976).

37. *Id*.

38. The 1976 Senate bill, S. 3219, did require the use of "best available control technology" (BACT) as a part of PSD requirements for sources located in clean air regions. *See* S. Rep. No. 717, 94th Cong., 2d Sess. 139, 158 (1976) (reprinting S. 3219). This BACT requirement, though, was keyed to the protection of clean air regions and did not require scrubbing on a national basis. Moreover, as originally conceived by EPA, BACT merely required use of any control strategy sufficient to meet existing NSPSs, including the use of either low-sulfur coal or scrubbers. 40 C.F.R. §§ 52.01(f), 52.21(d)(2)(ii) (1976). Since the passage of the 1977 amendments, EPA has not treated BACT as requiring standards more stringent than NSPS, even where "demonstrated" technologies are capable of higher percentage reductions than required by NSPS. *See, e.g.*, EPA, Compilation of BACT/LAER Determinations 14, 16 (1979) (EPA Pub. No. 450/2–79–003).

39. H.R. Rep. No. 1742, 94th Cong., 2d Sess. 89 (1976) (conference report).

40. *See* 7 Envir. Rep. (BNA) 835 (1976).

CHAPTER 4

1. *See, e.g., 1977 House Hearings, supra* note 35, chap. 2, at 1677–78 (statement of Douglas Costle, Administrator, EPA).

2. For an informal analysis of the management process that produced the National Energy Plan, see N.Y. Times, Apr. 24, 1977, § 1, at 1, col. 1.

3. *See* 7 ENVIR. REP. (BNA) 1633 (1977) (summary of briefing paper).

4. The president announced the National Energy Plan in April 1977. *See* Address to the Nation, 1 PUB. PAPERS 656 (Apr. 18, 1977); Address Delivered before a Joint Session of the Congress, *id.* at 663 (Apr. 20, 1977). The showpieces of the program were a series of taxes on consumers and producers of gasoline, some cautious steps toward deregulation of energy markets, and tax incentives and subsidies for energy conservation. *See* National Energy Program: Fact Sheet on the President's Program, *id.* at 672, 674–87 (Apr. 20, 1977).

5. *See 1977 House Hearings, supra* note 35, chap. 2, at 1677, 1678 (statement of Douglas Costle) (supporting use of high-sulfur coal and BACT); *Coal Conversion Legislation: Hearings before the Subcomm. on Energy Production and Supply of the Senate Comm. on Energy and Natural Resources*, 95th Cong., 1st Sess. 1392, 1393–94 (1977) (statement of Douglas Costle) (supporting BACT and PSD as mitigating effects of coal conversion); *id.* at 1407, 1412 (statement of John F. O'Leary, Administrator, Federal Energy Administration) (supporting BACT as compatible with energy policy).

Costle's statements can only be seen as the product of White House influence since expert opinion within the EPA had been moving in a very different direction. By late 1976, EPA analysts had concluded that universal scrubbing was an environmentally unsound policy. *See* J. Crenshaw, D. Kirchgessner, H. Kuo, C. Miranda, & A. Wehe, Staff Study, *supra* note 23, chap. 3, at i, 33 (concluding that lower emission ceiling was more economical and environmentally sound than a mandatory percentage reduction).

6. 33 CONGRESSIONAL QUARTERLY ALMANAC, 95TH CONGRESS, 1ST SESSION, 78–H to 80–H (list of House roll-call votes on Clean Air Act Amendments).

7. *See* Clean Air Act Amendments of 1977, § 1103A [*sic*], 123 CONG. REC. H5021 (daily ed. May 25, 1977) (proposal by Rep. Miller of Ohio). Rep. Miller's proposed amendment would have exempted power plants in a given state from the Clean Air Act whenever the state's

unemployment rate exceeded 6.5%. The amendment was defeated without a roll call. *Id.* at H5023.

8. 42 U.S.C. § 7425 (Supp. II 1978). Rep. Rogers's original amendment appears at 132 Cong. Rec. H5026–27 (daily ed. May 25, 1977).

9. *Id.* at H5027.

10. *Id.* at S9449 (daily ed. June 10, 1977) (amendment introduced on behalf of Sen. Metzenbaum (Ohio), Sen. Randolph (W. Va.), Sen. Bayh (Indiana), Sen. Heinz (Pa.)).

11. *Id.* at S9459.

12. *Id.* at S9468.

13. *See, e.g.*, EPA Preliminary Analysis, *supra* note 20, chap. 3, at V–5.

14. For a discussion of the costs of mandatory scrubbing, *see* pp. 66–74, 90, 102 *infra*.

15. *See* H.R. Rep. No. 564, 95th Cong., 1st Sess. 130, *reprinted in* [1977] U.S. Code Cong. & Ad. News 1510 (conference report; Senate did not change § 111). [hereinafter cited as 1977 Conference Report].

16. *See* 8 Envir. Rep. (BNA) 155–58 (1977).

17. *Id.* at 462.

18. *See id.* at 509 (approval of conference amendments); 123 Cong. Rec. H8672 (daily ed. Aug. 4, 1977) (House approval of conference bill); *id.* at S13,711 (Senate approval).

19. The description of Senator Domenici's actions is derived from interviews with aides to various senators and representatives on the House and Senate committees and the conference committee. Prior to passage of the Senate bill on June 10, Senator Domenici submitted additional statements criticizing the House's revision of Section 111. 123 Cong. Rec. S9477 (daily ed. June 10, 1977).

20. Interview with representatives of the Utility Air Regulatory Group in New Haven, Connecticut (Oct. 22, 1979). For evidence of the utility industry's early uncertainty over the actual impact of the 1976 House amendment to Section 111, *see Senate Hearings—1977 Amendments, supra* note 28, chap. 3, at 289–98 (analysis by George C. Freeman, discussing ambiguity of amendments to § 111 in 1976 Conference bill).

21. Section 111(h)(2) defines the phrase "not feasible to prescribe or enforce a standard of performance" to mean "any situation in which the Administrator determines that (A) a pollutant or pollutants cannot be emitted through a conveyance designed and constructed to emit or capture such pollutant . . . or (B) the application of measurement methodology to a particular class of sources is not practicable due to technological or economic limitations." Clean Air Act Amendments of

1977, § 111(h)(2), 42 U.S.C. § 7411(h)(2) (Supp. II 1978). Power plants emitting pollutants through a smokestack equipped with continuous monitors clearly fall outside this definition. This definitional amendment was also apparently intended to resolve uncertainties generated by Adamo Wrecking Co. v. United States, 432 U.S. 275, 285–89 (1978), which questioned the administrator's authority to issue work practice requirements as if they were "standards of performance."

22. Section 111, as amended in 1977, distinguishes between fossil-fired stationary sources (primarily power plants and industrial boilers), § 111(a)(1)(A), and all other sources, § 111(a)(1)(B). Percentage reduction requirements apply only to sources in the former category. Clean Air Act Amendments of 1977, § 111(a)(1)(A)–(B), 42 U.S.C. § 7411 (a)(1)(A)–(B) (Supp. II 1978).

23. Clean Air Act Amendments of 1977, § 111(a)(1)(A)(ii), 42 U.S.C. § 7411(a)(1)(A)(ii) (Supp. II 1978).

24. 42 U.S.C. § 7411(b)(2) (Supp. II 1978).

25. 42 U.S.C. § 7411(b)(1) (Supp. II 1978).

26. 42 U.S.C. § 7411(a)(7)(A) (Supp. II 1978).

27. 1977 CONFERENCE REPORT, *supra* note 15, at 130, [1977] U.S. CODE CONG. & AD. NEWS, at 1510.

28. *Id.* at 130, [1977] U.S. CODE CONG. & AD. NEWS, at 1511.

29. Clean Air Conference Report (1977): Statement of Intent; Clarification of Select Provisions, 123 CONG. REC. H8662–64, *reprinted in* [1977] U.S. CODE CONG. & AD. NEWS 1570–77.

30. *Id.*

31. For some insightful general remarks, *see* R. DAHL, AFTER THE REVOLUTION? 42–56 (1970) (time as valuable and scarce resource).

32. *See generally* N. POLSBY, CONGRESS AND THE PRESIDENCY 110 (3d ed. 1976) (increasing autonomy of specialized subcommittees); Ornstein, *Causes and Consequences of Congressional Change: Subcommittee Reforms in the House of Representatives, 1970–73*, in CONGRESS IN CHANGE 88, 105–10 (N. Ornstein ed. 1975) (reforms guaranteeing budgets, staff, and specific jurisdiction to subcommittees).

33. Between 1957 and 1976, staffing on congressional committees tripled, and congressmen's personal staffs grew by 180%. H. FOX & S. HAMMOND, CONGRESSIONAL STAFFS: THE INVISIBLE FORCE IN AMERICAN LAWMAKING 171 (1977). Committee staffing grew by 81% between 1972 and 1976 alone. *Id.*

34. *See generally id.* at 143–45; *Senate Committee System: Hearings before the Temporary Select Committee to Study the Senate Committee System*, 94th Cong., 2d Sess. 104 (1976) (statement of Sen. Morgan of N.C.) ("[A]

great deal of the legislation in this Congress is initiated by staff," because Senators "just do not have the time to read the records and to do
the necessary research that is necessary to bring forth meaningful and
important legislation.") For a carefully documented discussion of the
use and misuse of staff work by congressional subcommittees, *see* M.
MALBIN, UNELECTED REPRESENTATIVES: CONGRESSIONAL STAFF AND THE
FUTURE OF REPRESENTATIVE DEMOCRACY (1980).

CHAPTER 5

1. *See* 42 U.S.C. §§ 109–10 (Supp. II 1978). The regulation of old
plants is discussed further at 9–10 *supra*.

2. In the Northeast, composite urban average SO_2 emissions declined from 88 $\mu g/m^3$ in 1964 to 41 $\mu g/m^3$ in 1971. EPA, THE NATION-
AL AIR MONITORING PROGRAM: AIR QUALITY AND EMISSION TRENDS,
ANNUAL REPORT 4–17 to 4–19 (1973) (EPA Pub. No. 450/1–73–001–a)
[hereinafter cited as NATIONAL AIR MONITORING PROGRAM—1973].
Concentrations declined slightly between 1972 and 1977. EPA, NA-
TIONAL AIR QUALITY, MONITORING, AND EMISSIONS TRENDS REPORT 3–8
(1978). These declines in SO_2 *concentrations* occurred despite an estimated 45% increase in total nationwide *emissions* of SO_2 between 1960
and 1970, NATIONAL AIR MONITORING PROGRAM—1973, *supra*, at 4-3,
illustrating the uncertain connection between air quality and absolute
emission levels. The decline in concentrations of SO_2 is generally explained by reductions in residential and commercial coal burning in
urban areas, increased stack heights, and increased geographical dispersion of sources. *See* NATIONAL ACADEMY OF SCIENCES, SULFUR OX-
IDES 25 (1978) [hereinafter cited without cross-reference as NAS,
SULFUR OXIDES].

3. *See* U.S. COUNCIL ON ENVIRONMENTAL QUALITY, *supra* note 1, chap.
1, at 35, 45, 49 (figure 1–9). The most serious problems today occur in
areas affected by smelters, *id.* at 45, which can now defer compliance
with SO_2 standards until 1988. *See* 42 U.S.C. § 7419(c) (Supp. II 1978).

4. Many scientists have stated that there is no need to tighten present
ambient SO_2 standards. Cooper & Hamilton, Atmospheric Sulfates and
Mortality—The Phantom Connection, *reprinted in Oversight, OTA's
Study: The Direct Use of Coal: Hearing before the Subcomm. on Energy Devel-
opment and Applications of the House Comm. on Science and Technology*, 96th
Cong., 1st Sess. 72, 79 (1979) at 79 (standards probably too stringent);
Ferris, *Health Effects of Exposure to Low Levels of Regulated Air Pollutants: A
Critical Review*, 28 J. AIR POLLUTION CONTROL ASS'N 482, 493–94 (1978)

(standards "reasonable"); NAS, Sulfur Oxides at 166 (same). New plants meeting the pre-1977 NSPS generally do not cause increases sufficient to violate the regional ambient SO₂ standards, unless they are situated in close proximity to existing emission sources. *See* 1978 SO₂ Background Information, *supra* note 11, chap. 2, at 6–12 (percentage of ambient limits contributed by different sizes of power plants).

5. *See* R. Tobin, The Social Gamble: Determining Acceptable Levels of Air Quality 21 (1979); 1 U.S. Council on Environmental Quality, Annual Report 87 (1970) (monitoring systems "so spotty in coverage that it is very difficult to determine trends in the quality of air"). Combined federal and state monitoring systems are now far more extensive, with data recorded by more than 2,600 SO₂ monitors and 4,000 TSP (total suspended particulate) monitors, which are generally capable of measuring SO₄ concentrations. Even today, though, the quality of data produced by many systems is poor. *See* General Accounting Office, Report by the Comptroller General of the United States, Air Quality: Do We Really Know What It Is? 2, 5–17 (1979) (No. CED–79–4).

6. U.S. Dep't of Health, Educ. & Welfare, Air Quality Criteria for Sulfur Oxides 118–19 (1969) [hereinafter cited as 1969 Air Quality Criteria]. Few experts ever attributed the known effects of air pollution solely to SO₂. The air quality criteria document on which EPA's standard for SO₂ is based explicitly stated its findings in terms of sulfur oxides and *not* SO₂.

7. *See* 1969 Air Quality Criteria, *supra* note 6, at 119–26. In the October 1948 Donora incident, 20 people died, 10% of the town's population was severely affected and 43% was affected to some degree. The London fog in December 1952 resulted in 4,000 more deaths than would have been expected over a two-week period. Most deaths occurred among the elderly and persons with preexisting pulmonary or cardiac problems. No pollutant monitors were present in Donora during the incident; a single monitoring site in London recorded extraordinarily high daily concentrations of SO₂ and particulates. No measurements were made of particular species of particulates such as sulfates. Other episodes have occurred in the world's major cities since 1952, but none has been as serious. *See* Schimmel, *Evidence for Possible Acute Health Effects of Ambient Air Pollution from Time Series Analysis: Methodological Questions and Some New Results Based on New York City Daily Mortality, 1963–1976*, 54 Bull. N.Y. Acad. Med. 1052, 1054 (1978).

8. 1969 Air Quality Criteria, *supra* note 6, at 120–24. For a detailed account of the political and, to a lesser extent, scientific contro-

versies preceding the establishment of the first national ambient air quality criteria for sulfur oxides in 1967, *see* R. TOBIN, *supra* note 5, at 33–50 (stressing paucity of reliable scientific data).

9. 1969 AIR QUALITY CRITERIA, *supra* note 6, at 117. This problem of joint causation has continued to plague epidemiologists. *Cf.* Ferris, *supra* note 4, at 482, 491 (1978) (reporting study finding health benefits over decade in which particulate levels decreased and sulfur oxide levels increased).

10. NAS, SULFUR OXIDES 133, 166–71 (1978). *See also* Cooper & Hamilton, *supra* note 4, at 72, 79 (1979) (reporting respected experts' opinions that present SO_2 standards are unduly stringent). Even when subjects were briefly exposed to laboratory concentrations thirty or more times the primary ambient standard, acute effects were only barely detectable. 1969 AIR QUALITY CRITERIA, *supra* note 6, at 91–94. These results may be due to the fact that SO_2, a gas, is sufficiently water soluble that it tends to be absorbed in the respiratory passages. It is unlikely that more than 10% of inhaled SO_2 penetrates as far as the larynx. NAS, SULFUR OXIDES, at 134.

Still, some restrictions on peak levels of SO_2 may be necessary. Laboratory tests of SO_2 effects do not replicate ambient air conditions, and few tests have studied the effects of pollutants in combinations. *Id.* at 10. Other pollutants might aggravate the effects of SO_2. *Id.* at 153–55. Moreover, it is likely that at some point before ordinarily healthy people are affected, higher SO_2 concentrations will begin to affect asthmatics and other sensitive groups. *Id.* at 158.

11. *See* NAS, SULFUR OXIDES, at 143–53, 158–60; U.S. COUNCIL ON ENVIRONMENTAL QUALITY, *supra* note 1, chap. 1, at 58.

12. Several recent epidemiological studies have established statistically significant associations between concentrations of sulfates and regional mortality rates. *See, e.g.*, L. LAVE & E. SESKIN, AIR POLLUTION AND HUMAN HEALTH (1977); Mendelsohn & Orcutt, *An Empirical Analysis of Air Pollution Dose–Response Curves*, 6 J. ENVT'L ECON. & MANAGEMENT 85 (1979). Many epidemiologists question whether these new studies have adequately demonstrated that the observed effects are *caused* by the sulfates in the amounts measured. *See, e.g.*, NAS, SULFUR OXIDES, at 196–98 (assessment of Lave and Seskin); F. Lipfert, The Association of Human Mortality with Air Pollution 191–92, 196–97 (Ph.D. dissertation, Union Graduate School, Ohio, 1978) (analysis suggesting small but statistically significant association between mortality and general air pollution levels, but not between mortality and sulfates). One reason for doubt is that laboratory tests exposing human subjects for short periods

to sulfate concentrations more than twenty times the typical atmospheric concentrations have not shown adverse physiological reactions. *See* NAS, Sᴜʟꜰᴜʀ Oxɪᴅᴇꜱ, at 158–60. More generally, retrospective research of the sort carried out by Lave and Seskin cannot control for important confounding variables, such as variations in smoking habits among the populations studied. Therefore, many researchers evaluate the results as suggestive but inconclusive. *See* NAS, Sᴜʟꜰᴜʀ Oxɪᴅᴇꜱ, at 196–98; F. Lipfert, *supra*, at 182–86.

13. *See* NAS, Sᴜʟꜰᴜʀ Oxɪᴅᴇꜱ, at 40 (unreliability of historic data base); Forrest & Newman, *Ambient Air Monitoring for Sulfur Compounds: A Critical Review*, 23 J. ᴏꜰ Aɪʀ Pᴏʟʟᴜᴛɪᴏɴ Cᴏɴᴛʀᴏʟ Aꜱꜱ'ɴ 761 (1973) (same).

14. For a brief description of one such project, *see* Perhac, *Sulfate Regional Experiment in Northeastern United States: The 'SURE' Program*, 12 Aᴛᴍᴏꜱ. Eɴᴠɪʀ. 641 (1978). The SURE Program is operated by the Electric Power Research Institute, a utility-sponsored research organization.

15. *See* Ferris, *supra* note 4, at 493.

16. Nonurban sulfate concentrations in the northeastern United States have risen steadily since the early 1960s. NAS, Sᴜʟꜰᴜʀ Oxɪᴅᴇꜱ, at 21, 25 (based on extremely limited data). In contrast, urban concentrations, at least in the few locations where they were monitored, declined sharply during the initial period of urban cleanup in the 1960s. For example, sulfate concentrations in New York City declined from about 33 $\mu g/m^3$ in 1964 to 13 $\mu g/m^3$ in 1974. Eisenbud, *Levels of Exposure to Sulfur Oxides and Particulates in New York City and Their Sources*, 54 Bᴜʟʟ. N.Y. Aᴄᴀᴅ. Mᴇᴅ. 991, 1003 (1978). Since 1970, however, urban sulfate levels, as measured by the National Air Surveillance Network (NASN), have remained generally constant and nonurban levels have risen slightly, NAS, Sᴜʟꜰᴜʀ Oxɪᴅᴇꜱ, at 25, even though total emissions slightly declined. EPA, Mᴏɴɪᴛᴏʀɪɴɢ ᴀɴᴅ Aɪʀ Qᴜᴀʟɪᴛʏ Tʀᴇɴᴅꜱ Rᴇᴘᴏʀᴛ, 1974 125–26 (1976) (EPA Pub. No. 450/1–76–001) (1970 to 1974 estimates).

17. EPA, Aɪʀ Qᴜᴀʟɪᴛʏ ꜰᴏʀ Nᴏɴᴍᴇᴛᴀʟʟɪᴄ Iɴᴏʀɢᴀɴɪᴄ Iᴏɴꜱ 1971 ᴛʜʀᴏᴜɢʜ 1974: Fʀᴏᴍ ᴛʜᴇ Nᴀᴛɪᴏɴᴀʟ Aɪʀ Sᴜʀᴠᴇɪʟʟᴀɴᴄᴇ Nᴇᴛᴡᴏʀᴋ 55–70 (1977).

18. See Perhac, *supra* note 14, at 642 (citing sources).

19. *See* NAS, Sᴜʟꜰᴜʀ Oxɪᴅᴇꜱ, at 50–56 (describing modes of SO_2 deposition); EPA, Sᴜʟꜰᴀᴛᴇꜱ ɪɴ ᴛʜᴇ Aᴛᴍᴏꜱᴘʜᴇʀᴇ: A Pʀᴏɢʀᴇꜱꜱ Rᴇᴘᴏʀᴛ ᴏɴ Pʀᴏᴊᴇᴄᴛ MISTT 18–25 (1977) (EPA Pub. No. 600/7–77–921).

20. *See* 40 C.F.R. § 50.4 to 50.5 (1979) (national ambient air quality standards for SO_2). Although the PSD system for clean air areas some-

times serves as the binding constraint on the total amount of sulfur oxide produced by a plant, PSD has its greatest impact on areas with the
lowest levels of SO$_4$. EPA, PROTECTING VISIBILITY 1–6(a) (1979) (EPA
Pub. No. 450/5–79–008) (map depicting clean air areas) [hereinafter
cited as PROTECTING VISIBILITY].

21. Shortening a plant's smokestack should, in theory, reduce the
time sulfur oxides reside in the atmosphere, thereby reducing the
chances that SO$_2$ will be transformed into SO$_4$. See EPA, SULFATES IN
THE ATMOSPHERE: A PROGRESS REPORT ON PROJECT MISTT, supra note
19, at 24–25; NAS, SULFUR OXIDES, at 41–49 (describing transformation
process). Of course, attainable reductions in sulfate levels will vary with
local meteorological conditions and terrain. Moreover, shorter stacks
will increase local concentrations of other pollutants coming out of
smokestacks—notably nitrogen oxides and particulates. Thus, the final
policy on stack size must be set with a complex set of trade-offs in mind.

22. The legislative history of the 1977 amendments provides evidence of this congressional concern in its consideration of a number of
the act's new sections. See, e.g., H.R. REP. No. 294, 95th Cong., 1st Sess.
191 (1977) (NSPS); id. at 122–24 (PSD); id. at 83–84 (ICS); 42 U.S.C.
§ 7423 (limit to stack heights). For a good history of the public debate
concerning sulfates, see R. TOBIN, THE SOCIAL GAMBLE: DETERMINING
ACCEPTABLE LEVELS OF AIR QUALITY chaps. 8–9 (1979).

23. See note 12 supra.

24. NAS, SULFUR OXIDES, at 52–54.

25. See id. at 65 (reporting measured increase in acidity of lakes); id.
at 70–71 (effects on aquatic life).

26. Likens, Wright, Galloway, & Butler, Acid Rain, SCIENTIFIC AMER-
ICAN 43, 49–50 (Oct. 1974).

27. See, e.g., Likens & Bormann, Acid Rain: A Serious Regional Environmental Problem, 184 SCIENCE 1176, 1178 (1974); NAS, SULFUR OX-
IDES, at 80–82, 112–15.

28. Even if all Midwestern plants subject to the new (1979) NSPS
were allowed to emit at 1.2 pounds per MBTU, these plants would emit
only 11% of the sulfur oxides that will be produced by utilities in the
Midwest in 1995. ICF, INC., THE FINAL SET OF ANALYSES OF ALTER-
NATIVE NEW SOURCE PERFORMANCE STANDARDS FOR NEW COAL-FIRED
POWER PLANTS C–II–3a (June 1979) (prepared between December 1978
and May 1979). Overall, in 1995, new plants emitting SO$_2$ at a rate of
1.2 pounds per MBTU would produce 28% of the sulfur oxides produced by utilities east of the Mississippi. Id. at C–II–3a, 3b.

Although some debate exists, many scientists now believe that much of the sulfate concentration in the Northeast originates in the Midwest. (This book uses the term "midwestern" to describe power plants in the East North Central Region shown in map at p. 83, *infra* (Ohio, Indiana, Illinois, Michigan and Wisconsin)). *See, e.g.*, U.S. DEPT. OF ENERGY, REGIONAL ISSUE IDENTIFICATION AND ASSESSMENT: FIRST ANNUAL REPORT 1–10, 2–8, 3–7 (1979) (modeling results predicting that sulfur oxides transported from Midwest significantly affect SO_4 concentrations in New England, New York, and mid-Atlantic areas); Wolfe & Lioy, *Transport of Suspended Particulate Matter into the New York Metropolitan Area*, 54 BULL. N.Y. ACAD. MED. 1032–44 (1978) (suggesting that transport of SO_4 from Midwest associated with certain weather patterns significantly increases SO_4 levels in New York City area); *General Discussion: Session I*, 54 BULL. N.Y. ACAD. MED. 1045–51 (1978) (defense of Wolff & Lioy, *supra*, by one of its authors).

29. *See* J. Kilgroe & J. Strauss, Use of Coal Cleaning for Air Quality Management 2, 10 (table 3) (Jan. 22–25, 1980) (unpublished paper presented at Second Conference on Air Quality Management in the Electric Power Industry).

30. *See* J. Kilgroe, Coal Cleaning for Sulfur Oxide Emission Control 20 (Apr. 8–9, 1980) (unpublished paper presented at Acid Rain Conference, Springfield, Va.). Other authors have estimated that as little as a quarter of United States coal is presently washed. Bromley & Foley, *The Clean Fuel Supply: Factors Affecting U.S. and European SO_2 Emissions in the Mid-1980's*, in SYMPOSIUM ON COAL CLEANING, *supra* note 3, chap. 2, at 228, 234.

31. Estimates by Versar, Inc. (May 1980) (supplied to the authors by EPA). Versar reports that a reduction of 2.8 million tons a year is achievable by washing all cleanable steam coals mined in 1976 in major eastern and midwestern coal producing states. Since American utilities consumed 78% of available supplies of nonmetallurgical (steam) coal in 1976, ENERGY INFORMATION ADMINISTRATION, 2 ANNUAL REPORT TO CONGRESS 1978 at 91–93 (1978), 2.2 million tons of this reduction may be assigned to utilities. Because utilities classified as existing sources will consume roughly as much coal in 1995 as in 1975, ICF, Inc., *supra* note 28, at C–II–5, this same reduction from existing plants can be achieved at least as late as that year.

32. J. Kilgroe & J. Strauss, *supra* note 29, at 22–23. Note the wide range in marginal costs that prevails under both scrubbing and washing. Relatively large scrubbers have low marginal costs thanks to economies of scale. *Id.* at 23. Scrubbing costs also depend upon the sulfur content of the fuel burned and the percentage scrubbing required.

As far as washing is concerned, the marginal cost involved in washing high-sulfur coal tends to be lower than the marginal cost of washing lower sulfur varieties. This is because the costs involved in crushing and separating pyritic sulfur are not proportional to increases in the sulfur content of the pyrites. Thus washing 30% out of a 9-pound coal may generate three times the sulfur reduction achieved by washing 30% out of a 3-pound coal, without proportional increases in cost. *Id.* at 28. Moreover, it would appear that the Kilgroe and Strauss estimates understate washing's relative cost advantage. As Kilgroe and Strauss explain, washing coal tends to increase boiler reliability, decrease maintenance, and reduce the cost of fly ash removal. These nonenvironmental benefits reduce the net cost of the technology. *Id.* at table 5, 27. In contrast, scrubbing generates no similar savings in ordinary plant operation.

While Kilgroe and Strauss's optimistic assessment of coal washing seems to be supported by the studies they cite, there are studies that suggest a smaller cost advantage in favor of washing. For example, a recent TVA study suggests that the average cost of scrubbing at 85% is only slightly more expensive than the average cost of washing at about 35%. *See* J. Kilgroe, *supra* note 30, at 24–26. More generally, the data base on washing costs seems weak—reflecting the low institutional priority given the question until very recently. Similarly, the second-order environmental effects of disposing of the waste products generated by the variety of washing and scrubbing systems deserves more focused study.

Finally, the cost estimates presented here ignore the savings attainable by the substitution of lower sulfur coals as opposed to the application of treatment technologies.

33. J. Kilgroe, *supra* note 30, at 22, 28. At least one EPA consultant estimates a potential reduction exceeding two million tons a year. *See* Estimates by Versar, Inc., *supra* note 31.

34. *See* EPA, COST ANALYSIS OF LIME-BASED FLUE GAS DESULFURIZATION SYSTEMS FOR NEW 500-MW UTILITY BOILERS 4–24 to 4–32 (1979) (EPA Pub. No. 450/5–79–003) (estimates by PEDCo Environmental, Inc.)

35. ICF, INC., *supra* note 28, at C–VIII–21.

36. *Id.* (predicted 1995 reduction of 960,000 tons). This figure, moreover, may overstate the relief scrubbing affords the northeastern quadrant where sulfate levels are highest. The number includes a reduction of 350,000 tons obtained in the southeastern states of Alabama, Kentucky, Mississippi, and Tennessee—some of which lie far outside

the northeast quadrant, as we have defined it. *See* map on page 63, *supra*. The extent to which these sulfur oxides are transported to the northeast is not clear.

Before 1995, full scrubbing promises even smaller gains. Thus, EPA's consultants predict that in 1990 a 90% scrubbing requirement will generate an extra reduction of only 390,000 tons east of the Mississippi. *Id.* at C–VIII–21.

37. In fact, as long as ambient air quality standards for sulfates remain unpromulgated, EPA cannot require a state to reduce emissions from existing plants by an amount greater than is necessary to meet local SO_2 standards. *See* Train v. NRDC, 421 U.S. 60, 98 (1975) (nondiscretionary duty to approve SIP revisions as long as consistent with legally promulgated air quality targets). In recent years a number of utilities have responded to high fuel costs by obtaining variances that permit them to *increase* their emissions of sulfur oxide at the same time stricter standards are being applied to new plants.

38. *See* Fisher & Peterson, *The Environment in Economics: A Survey*, 14 J. OF ECON. LIT. 1, 12–16 (1976); W. BAUMOL & W. OATES, ECONOMICS, ENVIRONMENTAL POLICY, AND THE QUALITY OF LIFE 209–369 (1979). For reasons presented in B. ACKERMAN, S. ROSE-ACKERMAN, J. SAWYER, D. HENDERSON, THE UNCERTAIN SEARCH FOR ENVIRONMENTAL QUALITY 260–81 (1974), we believe that a system in which pollution rights are sold off to the highest bidder is superior to the more familiar effluent tax alternative. In particular, it is easy to modify the auction system to take into account the fact that new plants need to buy rights for longer periods of time than old plants require. *See id.* at 268.

39. To the extent that current operating problems experienced by scrubbers result mainly from insufficient attention to the design and operation of control technology, *cf.* note 42, *infra* (improper design and operation of control equipment found to be a major source of excess emissions), simple good faith efforts might represent a significant advance over current practices. EPA has long asserted that proper operation could significantly reduce the problems experienced by scrubbers. *See generally* FGD SYSTEMS CAPABILITIES, *supra* note 5, chap. 2, at 4–2 to 4–14.

40. *See* ENTROPY ENVIRONMENTALISTS, INC., AN EVALUATION OF THE CONTINUOUS MONITORING REQUIREMENTS OF THE SEPTEMBER 19, 1978 SUBPART DA NSPS PROPOSAL 33–38 (January 1979) (reporting 1978 studies showing monitors available for use only 51% to 57% of the period studied). The cited authority was submitted to the EPA Docket on behalf of the utility industry but relies in part on earlier studies done by

Entropy Environmentalists for EPA. *Id.* at 33–35. The EPA now generally agrees with Entropy's analysis of monitor reliability, and in its 1979 NSPS decision revised its proposed monitoring requirements in an attempt to take monitor failure into account. *See* 44 Fed. Reg. 33,580, 33,610 (1979).

41. *See* GENERAL ACCOUNTING OFFICE, REPORT BY THE COMPTROLLER GENERAL OF THE UNITED STATES: IMPROVEMENTS NEEDED IN CONTROLLING MAJOR AIR POLLUTION SOURCES 7–8, 11–13 (1979) (No. CED–78–165).

42. *See Complying Power Plants Exceed Air Limits by 25 Percent, Drayton Tells Chamber*, 11 ENVIR. REP. (BNA) 5 (1980) (reporting 71% of "complying" sources studied experienced incidents of excess emissions; excess emissions from plants not in compliance average 25% of the amount allowed by emission standards). EPA attributes roughly two-thirds of the excess emissions to either improper design or improper operation of pollution-control equipment. Interview with analyst, EPA Planning Office, in Washington, D.C. (June 25, 1980).

43. *See* 40 C.F.R. §§ 60.47a–60.49a (1979) (monitoring and reporting requirements for 1979 NSPS). Direct responsibility for the enforcement of NSPS may belong either to the EPA or to state enforcement agencies, depending on whether or not the EPA delegates its authority. *See* 42 U.S.C. § 7414(b) (Supp. II 1978) (authority to delegate enforcement procedures to states). The effectiveness of state enforcement systems varies widely. GENERAL ACCOUNTING OFFICE, *supra* note 41, at 8–9.

44. *See* FGD SYSTEM CAPABILITIES, *supra* note 5, chap. 2, at 4–2 to 4–9; *see* M. Maxwell, Sulfur Oxides Control Technology in Japan 21, 23–24 (Interagency Task Force Report, June 30, 1978) (attributing successful operation of scrubbers in Japan in part to careful operation and special training of staff); REPORT OF THE HEARING PANEL, *supra* note 4, chap. 2, at 29 (suggestion that American utilities have failed to hire sufficient qualified personnel).

45. Commentators have attributed the successful use of scrubbers in Japan to the strength of the Japanese enforcement program. The Japanese operate central research centers that are usually linked directly via telemetry to stations monitoring emissions from a major source. M. Maxwell, *supra* note 44, at 24.

46. Although the enforcement problem is far more tractable under a low-sulfur strategy, it still exists. Even coal from the same mine may vary substantially in sulfur content. For a discussion of regulatory problems created by sulfur variability, *see* Memorandum from Walter C.

Barber, Director, Office of Air Quality Planning and Standards, EPA, to Barbara Blum, Deputy Administrator, EPA (Dec. 6, 1979), *reprinted in* 10 ENVIR. REP. (BNA) 1872, 1873 (1980).

47. *See* B. ACKERMAN, SOCIAL JUSTICE IN THE LIBERAL STATE 111–13, 202–17 (1980).

48. *See* NAS, SULFUR OXIDES 46–51, 180–204. *See also* Ferris, *supra* note 4, at 493–94.

49. EPA, SULFATES RESEARCH APPROACH 3 (1977) (EPA Pub. No. 600/8–77–004). In addition to EPA's own programs, the Electric Power Research Institute (EPRI), a utility-sponsored research organization plans to spend approximately $6 million in 1981 on the atmospheric behavior and effects of sulfur oxide. Information supplied by EPRI (August 18, 1980). The Department of Energy will also be spending $3 million in 1981. Information supplied by U.S. Dept. of Energy, Office of Environment (August 21, 1980). Finally, the Energy Security Act of 1980, Pub. L. No. 96–294, 94 Stat. 611, § 706, provides up to $50 million for a ten-year study of acid rain. It is not yet clear how much control the EPA will have over the course of this study.

The multibillion-dollar cost of forced scrubbing is discussed at length in the next chapter.

50. *See* Rall, *Review of the Health Effects of Sulfur Oxides*, 8 ENVT'L HEALTH PERSPECTIVES 97, 117 (1974) (plan for distinguishing effects of various pollutants); STATIONARY SOURCE EMISSION CONTROL, *supra* note 39, chap. 1, at 148 (stressing value of reducing present uncertainty).

51. *See* ICF, INC., STILL FURTHER ANALYSES OF ALTERNATIVE NEW SOURCE PERFORMANCE STANDARDS FOR NEW COAL-FIRED POWERPLANTS B–15 (January 1979) (estimating cost of scrubber for 500 MW plant at $56 million).

52. No formal studies on the cost of planned retrofitting exists. In an interview, an engineer active in the field of scrubber research expressed confidence that as long as space is provided for potential retrofitting, the extra costs of later installation will be relatively modest.

53. Both government and private industry are currently engaged in research to improve sulfur removal technologies. *See generally* S. Gage, *Remarks* in FGD SYMPOSIUM, *supra* note 5, chap. 2, at 2, 3–7; Laseke and Devitt, *Status of Flue Gas Desulfurization Systems in the United States*, in FGD SYMPOSIUM, *supra* note 5, chap. 2, at 22, 43–49; and Morasky and Dalton, *EPRI's Flue Gas Desulfurization Program, Results, and Current Work*, in FGD SYMPOSIUM, *supra* note 5, chap. 2, at 96, 96–117. Indeed, EPA officials are already expressing optimism over the development of new, improved technologies that are readily retrofittable. *See EPA Says*

New Technology May Provide Much Cheaper Sulfur Dioxide Removal, 11 ENVIR. REP. (BNA) 63–64 (1980) (development of "exciting" new technology that can be retrofitted for 20% of present cost of scrubber retrofit).

54. EPA consultants now predict that emissions east of the Mississippi will decline from 16.0 million tons in 1975 to 13.1 million tons in 2010. *See* ICF, INC., *supra* note 28, at C–II–3a–b, D–1a.

55. *See* Ball & Matheny, Impediments to Air Quality Control: The Problem of Long Range Transport, Proceedings, Institute of Environmental Science, figure 4 (n.d.) (ton of sulfur dioxide produced in heavily industrialized areas of Midwest may expose two to five times as many people to given concentration of sulfate as same amount emitted in most areas of New England). In some areas, such as New England, short distances separate sites with large differences in population exposure. *Id*. For an excellent study of control alternatives in one metropolitan area *see* R. MENDELSOHN, TOWARDS EFFICIENT REGULATION OF AIR POLLUTION FROM COAL-FIRED POWER PLANTS 138–60, 177 (1978) (concluding that siting considerations minimize cost of abatement on local level under certain conditions).

56. *See* EPA, SULFATES IN THE ATMOSPHERE: A PROGRESS REPORT ON PROJECT MISTT, *supra* note 19, at 24–25. While the EPA is presently revising its policy on stack height, it is too soon to guess the extent to which the agency will squarely confront the complex policy trade-offs involved. *See* p. 64 and note 21, *supra*.

57. *See* U.S. DEPT. OF ENERGY, FEDERAL ENERGY REGULATORY COMMISSION, STATUS OF COAL SUPPLY CONTRACTS FOR NEW ELECTRIC GENERATING UNITS: 1977–1986 at 3 (Supp. 1978).

58. Due to the effects of sulfur variability on the regulation of power plants, *see* Memorandum from Walter C. Barber to Barbara Blum, *supra* note 46, EPA has recently begun extensive work on the problems of predicting both the long- and short-term average sulfur content of known coal reserves. *See* J. Kilgroe & J. Strauss, *supra* note 29, at 21–22. (also citing sources). Given the size of America's coal reserves, such work is only beginning. Nonetheless, some progress is being made. For example, where variability is a problem it now appears that coal cleaning may not only reduce average sulfur content but its associated variability as well.

59. Sulfur oxides affect visibility by reducing visual range. A more complete discussion would include other types of visibility impairment caused by other pollutants—e.g., the discoloration of the air caused by nitrogen oxides.

Visual range is usually defined as the distance at which a typical observer under daytime conditions can distinguish a target on the horizon with a visual contrast of 2% (the theoretical contrast limit of the human eye). Experimental data have suggested, however, that the typical human eye may not detect contrasts smaller than 5%. A. STERN, 2 AIR POLLUTION 14, 17–18 (3d ed. 1977); see W. MIDDLETON, VISION THROUGH THE ATMOSPHERE 92–95 (1952). The uncertainty introduced by this and other assumptions is substantial. Nevertheless, a 2% limit for "standard" observers is commonly assumed by researchers. See PROTECTING VISIBILITY, supra note 20, at 8–10.

60. Visual range can be calculated from the formula $V_r = 3.92/b_{ext}$, where V_r is in kilometers and b_{ext} is a light extinction coefficient expressed in inverse kilometers (km^{-1}). In particle-free air, b_{ext} has a value of approximately 0.12 km^{-1}. (The calculations in the text are based on the assumption that each additional $\mu g/m^3$ of SO_4 increases b_{ext} by $0.04 \times 10^{-4} m^{-1}/\mu g/m^3$.) In perfectly clean air, visual range is so great that a precise calculation of its value must take into account complications introduced by the curvature of the Earth. The figures appearing in the text ignore these complications. For a more complete treatment, see PROTECTING VISIBILITY, supra note 20, at 2–18 to 2–23. Average visual range in the East is less than 10 miles at many times and places. Id. at 4–26 (figure 4–8).

61. EPA, VISIBILITY IN THE SOUTHWEST: AN EXPLORATION OF THE HISTORICAL DATA BASE 37 (1978) (EPA Pub. No. 600/3–78–039) [hereinafter cited as VISIBILITY IN THE SOUTHWEST].

62. See generally PROTECTING VISIBILITY, supra note 20, at 5–1 to 5–4.

63. See PROTECTING VISIBILITY, supra note 20, 5–10 to 5–11. Other pollutants—especially nitrogen oxides—produce far more noticeable plume effects within 30 miles of the plant than sulfur oxides. Id. at 5–11 to 5–13.

64. Currently power plant compliance is assessed on the basis of the plant's annual average discharge, with statistical adjustments to prevent violation of short-term ambient air standards. Memorandum from Walter C. Barber to Barbara Blum, supra note 46. The EPA proposes to enforce emission limitations for its new NSPS using monthly averages. 40 C.F.R. § 60.43a(c) (1979). In the past, EPA prepared regulations more sensitive to shifting meteorological conditions that were tentatively endorsed by the National Academy of Sciences, see STATIONARY SOURCE EMISSION CONTROL, supra note 39, chap. 1, at 216.

65. Models needed for predicting plume effects at sufficiently long

distances have yet to overcome a number of technical problems, and regional-scale models have yet to be validated. *See* PROTECTING VISIBILITY, *supra* note 20, at 5–20 to 5–24. Current EPA thinking suggests that for all but the largest sources, a power plant's impact on visual range in clean air areas is probably more sensitive to the plant's choice of site than to its emission rate. *Id*. at 5–14.

66. During a nine-month copper strike that closed western smelters, sulfate levels in Arizona dropped by 60% to 67%, and visibility improved across the state by 8% to 29%, *see* VISIBILITY IN THE SOUTHWEST, *supra* note 61, at 88–89, 92–93. In the eight-state Rocky Mountain Region smelters in 1976 emitted 2,484,000 tons of SO_2, compared to 430,000 tons emitted by power plants. EPA, 1976 NATIONAL EMISSIONS REPORT 15, 40, 78, 187, 198, 211, 318, 360 (1979) (EPA Pub. No. 450/4–79–019).

67. *See* Memorandum from John Bachmann (Air Office) to Walter C. Barber (Air Office) (Mar. 7, 1979) (EPA Docket No. OAQPS–78–I, Item No. II–B–37) (figure 1, table 1) (difference in emissions between partial and full scrubbing equal to 300,000 tons in 1995, would have no perceptible effect on visual range region-wide); *id*. at 4–5 (tentative findings as to overall improvement).

68. Memorandum from Walter C. Barber (Air Office) to David G. Hawkins (Air Office) at 1 (Apr. 12, 1979) (EPA Docket No. OAQPS–78–I, Item No. II–B–37). This supplemental analysis predicted that use of partial rather than full scrubbing would produce perceptible differences in visual contrast by the year 2010, but only because it was assumed that smelters would achieve major cutbacks by that date. *Id*. at 2 (for worst case conditions, and assuming large reductions in smelter emissions).

69. *See* 42 U.S.C. § 7419(c) (Supp. II 1978).

70. *See* note 59 *supra*.

71. *See* PROTECTING VISIBILITY, *supra* note 20, at 2–13 to 2–27, 5–20 to 5–26 (detailing areas requiring further research). We stress that our criticisms are largely based upon conventional beliefs prevailing among EPA experts on visibility. Only in 1977, though, did Congress explicitly direct the agency to apply the information it possesses and establish regulations specifically designed to protect visibility. Since the agency is still working on these regulations, it would be premature to comment upon them. The only thing that is clear is that it is such efforts in instrumental rationality that promise lasting environmental gains over the long run.

CHAPTER 6

1. Although the authors attempted to discuss NSPS with Mr. Hawkins during May of 1978, this proved impossible. The description of Air Office perspectives is based on other interviews and documents available in the EPA docket.

2. 42 U.S.C. § 7411(a)(1)(C) (Supp. II 1978).

3. *See* EPA, Draft Standards of Performance for New Stationary Sources: Electric Utility Steam Generating Units (EPA Docket No. OAQPS–78–I, Item No. II–C–194 (Nov. 29, 1977)).

4. The results of the modeling effort were reported in four separate documents issued by ICF, the agency's consulting firm, at different stages in the development of the rule-making proceeding: EFFECTS OF ALTERNATIVE PERFORMANCE STANDARDS FOR COAL-FIRED UTILITY BOILERS ON COAL MARKETS AND ON UTILITY CAPACITY EXPANSION PLANS (Sept. 1978) (prepared between October 1977 and April 1978) [hereinafter cited as ICF I]; FURTHER ANALYSES OF ALTERNATIVE NEW SOURCE PERFORMANCE STANDARDS FOR NEW COAL-FIRED POWER PLANTS (Sept. 1978) [hereinafter cited as ICF II]; STILL FURTHER ANALYSES OF ALTERNATIVE NEW SOURCE PERFORMANCE STANDARDS FOR NEW COAL-FIRED POWER PLANTS (Jan. 1979) (prepared between September and December 1978) [hereinafter cited as ICF III]; and THE FINAL SET OF ANALYSES OF ALTERNATIVE NEW SOURCE PERFORMANCE STANDARDS FOR NEW COAL-FIRED POWER PLANTS (June 1979) (prepared between December 1978 and May 1979) [hereinafter cited as ICF IV]. Some results from ICF I and ICF II appeared after further analysis in ELECTRIC UTILITY STEAM GENERATING UNITS: BACKGROUND INFORMATION FOR PROPOSED SO$_2$ EMISSION STANDARDS 2–1 to 3–28 (Supp. 1978) (EPA Pub. No. 450/2–78–007a–1) [hereinafter cited as 1978 SO$_2$ BACKGROUND INFORMATION (Supp.)] and in EPA's September 1978 proposal for SO$_2$ emission limits, 43 Fed. Reg. 42,165–68 (1978). Preliminary results from ICF III appeared in a supplemental notice, 43 Fed. Reg. 57,857–59 (1978), immediately prior to public hearings held on the proposal in December 1978. A summary of the results of ICF IV is reprinted in EPA's publication of its decision, 44 Fed. Reg. 33,608–09 (1979).

The model's predictions varied from analysis to analysis but the trends from the various reports generally remained constant. For example, all the studies show the same general pattern of regional emissions. To simplify the reader's task in following the decision-making process, we have consistently used figures from the final analysis, even though ICF IV was not available to decision makers until April of 1979. Where this ahistorical simplification significantly distorts the description of events, we refer to earlier analyses.

5. *See, e.g.* 1978 SO$_2$ BACKGROUND INFORMATION, *supra* note 11, chap. 2, 6–11 to 6–13 (analyzing "air quality impact" of new NSPS solely in terms of local SO$_2$ concentrations); 44 Fed. Reg. 33,605–07 (1979) (justifying 1979 NSPS simply in terms of reductions in nationwide SO$_2$ emissions). For an example of agency analysts' recognition of the extent to which a concern with sulfates might confound a regulatory scheme focused exclusively on SO$_2$ concentrations, *see* POSITION PAPER, *supra* note 30, chap. 2, at xii–xiv, xvi–xvii.

6. *See* ICF III, *supra* note 4, at 15 (figure 1). The discussion that follows assumes that concentrations of SO$_4$ are positively correlated, albeit to an unknown degree, with emissions of SO$_2$ from neighboring areas. By aggregating emissions in northern Kentucky—on the southern bank of the Ohio River—with emissions from states as far south as Alabama and Mississippi, the model tends to obscure, rather than refine, the relationship between SO$_2$ emissions and SO$_4$ concentrations in the Northeast.

7. Environmental groups were especially critical of the modelers' assumption that utilities would stretch out the lives of old plants when faced with increased costs for new plants. Natural Resources Defense Council & Environmental Defense Fund, Comments on Proposed Standard of Performance for New Electric Utility Steam Generating Units, at V–3 to V–6 (1978) (EPA Docket No. OAQPS–78–I, Item No. IV–D–631) [hereinafter cited as NRDC & EDF Comments]. This is admittedly a crucial assumption. We think it is a reasonable one.

8. The agency itself considered no standard requiring utilities to scrub less than 33% of the sulfur from their coal. *See* ICF III, *supra* note 4, at 12–17; ICF IV, *supra* note 4, at 3–4. The Utility Air Regulatory Group advocated a standard that would have allowed as little as 21% removal on some coals. *See* National Economic Research Associates, Inc., Comments on the Economic Impacts of EPA's September 19, 1978 Proposed Revision to New Source Performance Standards for Electric Utility Steam Generating Units 2 (Jan. 12, 1979) (EPA Doc. No. OAQPS–78–I, Item No. IV–D–611 app. D) (discussing UARG proposal). Even they considered this option in connection with relatively uncommon, low-sulfur coals (those containing less than 0.8 lb/MBTU of SO$_2$). *See* ICF III, *supra* note 4, at B–21. For more common coal types (containing more than 1.2 lb/MBTU), all "partial scrubbing" options consistently assumed at least 33% scrubbing. *See id.* at B–13 to B–21.

9. Full scrubbing promised to increase oil consumption by roughly 250,000 barrels a day compared to most other compliance strategies, ICF IV, *supra* note 4, at 5—or roughly 1% of current American consumption. This projected increase is generated largely by the model's assumption that the utilities would extend the lives of existing oil and coal burning plants when faced with the extra costs of scrubbing new plants.

10. *See* ICF IV, *supra* note 4, at C–VIII–21. Earlier modeling runs produced somewhat more optimistic estimates. Results published prior to the EPA's September 1978 proposal predicted nationwide decreases of 21%. *See* 1978 SO$_2$ BACKGROUND INFORMATION (Supp.), *supra* note 4, at 3–5 (table 3–2).

11. *See* ICF IV, *supra* note 4, at C–II–3a, C–VIII–3a (predicting 3% increase in emission tonnages in East North Central Region by 1995). The model's East North Central region includes Illinois, Indiana, Michigan, Ohio and Wisconsin. When speaking of air quality, we have referred to this region as the Midwest. When coal-producing regions are described, Ohio is part of the Appalachian Region, not part of the Midwestern Region. The EPA, in its September 1978 notice, used a slightly different definition of "midwestern," resulting in a slight decrease in predicted emission from the "Midwest." 43 Fed. Reg. 42,165 (1978).

12. *See* ICF IV, *supra* note 4, at 5. In contrast to the final figure of $4.1 billion, earlier computer runs had predicted that full scrubbing would "only" cost an extra $2.6 billion in 1995 when compared to the "old" NSPS. 43 Fed. Reg. 42,168 (1978) (table 8). Cost figures here and elsewhere are incremental annualized costs relative to the 1.2 standard. They include annualized capital costs as well as operating and maintenance costs. These figures do not represent the total annualized investment that utilities will make for control of sulfur oxides in 1995, since the costs necessary to reach the old 1.2 standard imposed in 1971 are not included. Other consultants have predicted that continued use of the 1971 NSPS would cost about $6 billion in 1990 (in annualized 1977 dollars). *See* 1978 SO$_2$ BACKGROUND INFORMATION (Supp.), *supra* note 4, at 3–38.

13. Clean Air Act Amendments of 1977, Pub. L. No. 95–95, § 109(a), 91 Stat. 685, 697 (codified at 42 U.S.C. § 7411(f)(1) (Supp. II 1978)).

14. Sierra Club v. Costle, No. 76–1297 (D.D.C., Aug. 25, 1978) (stipulation) (EPA Docket No. IV–J–25); *see EPA to Propose New Source Standard September 12, Hawkins Tells House Panel*, 9 ENVIR. REP. (BNA) 543, 544 (1978).

15. *Cf.* Memorandum from Walter C. Barber (Air Office) to Roy N. Gamse (Planning Office) (Aug. 9, 1978) (stating that a serious study of the benefits of NSPS would be "counterproductive.")

16. EPA routinely submitted its proposals to other executive offices for interagency review. *See, e.g.,* Letter from Don R. Goodwin (Air Office) to Addressees in other Executive Agencies (Dec. 1, 1977) (EPA Docket No. OAQPS–78–I, Item No. II–F–1) (enclosure of draft NSPS provisions).

17. *See, e.g.,* 9 ENVIR. REP. (BNA) 436–37 (1978) (summarizing DOE opposition); *id.* at 660 (reviewing letter from DOE Deputy Secretary John F. O'Leary, to Costle, favoring sliding-scale emission standard).

See also Letter from John O'Leary (DOE) to Douglas C. Costle (EPA) (July 6, 1978) (EPA Docket No. OAQPS–78–I, Item No. II–F–4); Letter from John F. O'Leary to Douglas C. Costle (Aug. 11, 1978) (EPA Docket No. OAQPS–78–I, Item No. II–F–5).

18. *See* 9 ENVIR. REP. (BNA) (1978); *id.* at 860 (reprinting memorandum).

19. The September notice actually proposed a 1.2 ceiling with certain exemptions and scrubber performance averaging 85%, measured on a *daily* basis. *Id.* at 42, 158. Most policy options considered after September 1978 were formulated in terms of monthly or yearly averages. EPA analysts we interviewed stated that they had generally considered the September notice's 1.2 standard (with exemptions) to be roughly equivalent to a 0.8–1.0 lb annual average. During the modeling process, a requirement that a scrubber reduce emissions by an average of 85% every day was generally considered to be equivalent to a requirement that it reduce emissions by an average of 90% over a longer period such as a month or year. *See, e.g.,* ICF II, *supra* note 4, at A2. For the reader's convenience, the authors have translated percentage reduction requirements into equivalent long-term averages throughout. Where the text refers to 90% scrubbing, long-term averages should be assumed unless otherwise noted.

20. 43 Fed. Reg. 42,154 (1978). The Janus-faced way EPA announced its continuing uncertainty in the *Federal Register* is explicable in terms of yet another aspect of the "new" § 111. The agency-forcing statute made the revised NSPS applicable from the date of its proposal, 42 U.S.C. § 7411(a)(2) (Supp. II 1978). *See* 9 ENVIR. REP. (BNA) 861 n. 1 (1978). Hence by formally proposing its most costly standard—full scrubbing—the agency preserved maximum flexibility. If it chose to stick with the Air Office proposal, the new NSPS would apply as of the September notice; if it chose a cheaper option, none of the utilities

would complain of the unfairness of the September date. In contrast, if the agency had merely announced that it had not made up its mind, complaints about retroactivity would be predictable if the EPA later imposed a harsh standard on all power plants proposed after its rule-making proceeding commenced.

21. *See* 9 ENVIR. REP. (BNA) 861 (1978) (reprinting Costle memorandum declaring that September notice will seek further information to help EPA make decision).

22. *See* ICF IV, *supra* note 4, at 5. An earlier set of computer analyses, completed by December 1978, contained a somewhat more diverse group of program options. Some of these proposals would have cut as much as 5% off the 1995 nationwide loading projected for full scrubbing. *See* ICF III, *supra* note 4, at 37.

23. Because partial scrubbing options reduced the cost of new plants, partial scrubbing could also produce net reductions in emissions by encouraging the earlier replacement of dirty old plants by new clean ones. *See* note 7, *supra*.

24. Nationwide coal production was expected almost to triple between 1975 and 1995. Even without forced scrubbing, ICF predicted that production would increase over the twenty years by 25% in Appalachia, by 190% in the Midwest, and by 750% west of the Mississippi, ICF IV, *supra* note 4, at C–II–9.

25. There are relatively few existing coal burners in the West, so this factor is not nearly as important as it is in some eastern regions. *See* ICF IV, *supra* note 4, at C–VIII–7a (utility coal consumption predicted to increase from 1.1×10^{15} BTU to 9.1×10^{15} BTU in states west of the Mississippi between 1975 amd 1995).

26. The overall reduction is modest because smelters, not power plants, contribute most of the region's SO_2. *See supra* note 66, chap. 5. ICF found that, at most, some sliding scales might add from 300,000 to 500,000 tons to the overall load in the Rocky Mountain and Pacific regions, when compared to full scrubbing. *See* ICF III, *supra* note 4, at D–III–2; and see map on p. 83 for model's definition of relevant regions. While such regionwide figures are too aggregative to be very revealing as to the effects on western visibility, see pp. 75–78, the Air Office concluded, after discussions with outside consultants, that changes of this order of magnitude will not make a perceptible difference. *See supra* notes 66–67, chap. 5.

27. In ICF III, *supra* note 4, the EPA modeled four different approaches that contemplated different treatment for different parts of the country. Three of these approaches reduced both costs and emis-

sions relative to forced nationwide scrubbing. The fourth option yielded lower costs without reducing emissions. The overall emission reductions promised by regional standards were generally small—on the order of 500,000 tons nationwide (or 2%–3% of all emissions). Predicted cost savings ranged from zero to $600 million per year by 1995. ICF III, *supra* note 4, at 37, D–III–3.

28. The Planning Office assumed that compliance with its proposed 0.55-lb. standard would be measured by calculating a plant's average emissions on an *annual* basis. For the importance of averaging periods, *see* note 19, *supra*.

29. ICF III, *supra* note 4, at 37.

30. *See* ICF III, *supra* note 4, at D–III–1 to –2. The eastern loadings reported in the text include the contribution of the model's East South Central Region—embracing the states of Alabama, Kentucky, Mississippi, and Tennessee. (See the map on page 83, *supra*.) Some of these states are remote from the Northeastern region that bears the brunt of the health and ecological risk generated by sulfates. (See the map on page 63, *supra*.) If emissions from the East South Central region are excluded from the analysis, the low ceiling reduces 1995 emissions in the Northeast by only 540,000 tons, and not the 740,000 tons reported in the text.

One should also retain perspective on the 280,000-ton increase in western power plant emissions by recalling that smelting contributed 2,484,000 tons of SO_2 in the eight-state Rocky Mountain region in 1976. *See supra* note 66, chap. 5. And given the smelting lobby's demonstrated capacity to delay compliance, see pp. 77–78, *supra*, the extent this load will be reduced over time remains unclear. For EPA predictions, *see* note 46 *infra*.

31. *See* Exec. Order No. 12,044, 3 C.F.R. 152 (1979). "Significant" regulations are those that impose costs of $100 million annually, or that have major impacts on individual industries, levels of local government, or geographic regions. *Id*. at § 3(a)(1), 3 C.F.R., at 154.

RARG is chaired by the chairman of the Council of Economic Advisers (CEA) and includes economic and regulatory members. Representatives of CEA, the Office of Management and Budget (OMB), and the Departments of Commerce, Labor, and Treasury constitute the economic group. Regulatory members come from the Departments of Health, Education and Welfare, Housing and Urban Development, Interior, Justice, Transportation, and the EPA. The Office of Science and Technology Policy participates in RARG as a member, and the president's Domestic Policy Staff and the Council on Environmental

Quality (CEQ) participate as advisers. *See* Procedures of the Regulatory
Analysis Review Group (Nov. 1978) (unsigned memorandum).

32. RARG reports are formally submitted by the Council on Wage
and Price Stability (COWPS), which operates with explicit statutory authority to intervene in the rule-making proceedings of agencies in order
to present its views on the inflationary impact of decisions. *See* 12
U.S.C. § 1904, § 3(a)(8) (Supp. II 1978). Presumably reliance on
COWPS is intended to minimize questions of the improper use of executive power on regulatory decisions.

33. *See* pp. 86–87, *supra*.

34. *See* Letter from George C. Freeman, Jr. to Stuart Eizenstat (Aug.
18, 1978) (EPA Docket No. OAQPS–78–I, Item No. II–D–433)
(restating concern expressed at July 13 meeting with Eizenstat); Letter
from Barbara Blum, Deputy Administrator, EPA, to James Schlesinger
(July 6, 1978) (EPA Docket No. OAQPS–78–I, Item No. II–F–7)
(referring to meeting on NSPS with the president, Schlesinger, and the
Business Roundtable).

35. Environmental Protection Agency's Proposal for the Revision of
New Source Performance Standards for Electric Utility Steam
Generating Units, Report of the Regulatory Analysis Review Group,
submitted by the Council on Wage and Price Stability (Jan. 15, 1979)
(EPA Docket No. OAQPS–78–I, Item No. IV–H–12) [hereinafter cited
as *RARG Report*.]

36. For a fuller discussion of the role of such reviewing institutions,
see B. ACKERMAN, S. ROSE-ACKERMAN, J. SAWYER, & D. HENDERSON, THE
UNCERTAIN SEARCH FOR ENVIRONMENTAL QUALITY 156–61 (1974).

37. *See RARG Report*, *supra* note 35, at 29.

38. Using its exposure index, RARG showed that, as a result of regional variations in emissions, some standards with lower nationwide
emissions would actually expose more people to high sulfur oxide concentrations than alternative standards with higher nationwide emissions. *RARG Report*, *supra* note 35, charts 1–3. As RARG itself admitted,
its exposure index was crude and very imprecise. This was hardly
RARG's fault however. During the summer of 1978, representatives of
the Council of Economic Advisers requested that the EPA conduct a
benefits study as part of the NSPS rule making. The Air Office opposed committing resources to such a study, calling it "counterproductive." *See* Memorandum from Walter C. Barber (Air Office) to
Roy N. Gamse (Planning Office) (Aug. 9, 1978).

39. *RARG Report*, *supra* note 35, at 2 (emphasis in original).

40. *See* note 38, *supra*.

41. In December 1978, ICF predicted that a 0.55 full scrubbing program would yield 500,000 tons less in 1995 nationwide emissions than the 0.55 partial scrubbing option and would increase annualized costs by $1 billion. Almost all of the decrease in emissions would occur west of the Mississippi. *See* ICF IV, *supra* note 4, at D–III–1.

42. Or so the Planning Office charged in a memorandum criticizing Air Office projections. *See* Memorandum from Roy N. Gamse (Planning Office) to Walter C. Barber (Air Office) (Feb. 28, 1979).

43. *See* ICF IV, *supra* note 4, at D–1a.

44. *See* Memorandum from Roy N. Gamse to Walter C. Barber, *supra* note 42.

45. *See* ICF IV, *supra* note 4, at D–1a.

46. The Planning Office estimated that in 1975 smelters emitted 2.2 million tons in the Rocky Mountain and Pacific regions, and power plants emitted only 0.6 million tons. *See* Memorandum from David W. Tunderman (Planning Office) to William Drayton 3 (Mar. 9, 1979) (figure 1); *but see* data reported *supra* at note 30. By 1995, the Planning Office expected smelter emissions to decline by an amount falling between 1.1 and 1.6 million tons. Under a 0.55 standard, SO_2 emissions from utilities in the far western region should increase by only 0.6 million tons between now and 2010—suggesting there will be little change in emissions in the Rocky Mountain and Pacific states. *See id.* (showing smelter emissions decline); ICF IV, *supra* note 4, at D–1a (showing increase in SO_2 emissions).

In contrast to the far West, the model's Central West regions, *see* map on page 83, *supra*, may experience an increase in emissions of an additional 800,000 tons, assuming partial scrubbing and no NSPS revisions. *See* ICF IV, *supra* note 4, at D–1a.

47. *See* Memorandum from Walter C. Barber (Air Office) to the file (undated) (EPA Docket No. OAQPS–78–I, Item No. IV–E–16) (report of EPA briefing held for Office of Management and Budget and the Departments of Agriculture and Interior, Feb. 16, 1979).

48. *See id.*, attachments D, E (summarizing briefing of Interior and other departments on NSPS and attaching Interior material on visibility problem).

49. *See* Memorandum from Walter C. Barber to the file (May 1979) (EPA Docket No. OAQPS–78–I, Item No. IV–E–18) (summarizing meeting).

50. *See* p. 76, *supra*.

51. *See* note 46, *supra*, and pp. 75–78 *supra*. *See also* PROTECTING VISIBILITY, *supra* note 20, chap. 5, at 6–5 to 6–7 (discussing the contribution of windblown dust to visibility impairment).

52. *See* Memorandum from Strategies and Air Standards Division to Walter C. Barber (Air Office) (March 7, 1979) (EPA Docket No. OAQPS–78–I, Item No. IV–B–36). A consensus of experts contacted by EPA initially disagreed on whether there would be a perceptible regionwide difference in visibility under full or partial scrubbing. Supplemental analyses done by EPA concluded that "emission differences between options in 1995 are too small to predict impacts." Memorandum from Walter C. Barber to David G. Hawkins (Apr. 12, 1979) (EPA Docket No. OAQPS–78–I, Item No. IV–B–87). The experts themselves stressed the tentative nature of their findings, and, in ways that were at times contradictory, suggested important factors that required further analysis.

53. *See* Memorandum from Walter C. Barber (Air Office) to the file (May 24, 1979) (EPA Docket No. OAQPS–78–1, Item No. IV–E–20). The description in the text here, as elsewhere, derives in part from interviews and not only the description found in the docketed memorandum that recorded the meeting.

54. We do not know whether the pictures displayed at the Costle press conference were the same as the discredited pictures presented by the Department of the Interior.

55. Old plants (including plants converted from oil) and plants regulated by the 1971 NSPS will together require 40% more coal in 1995 than power plants consumed in 1975. ICF IV, *supra* note 4, at C–II–15. In addition, high sulfur producers may well benefit from a possible boom in export sales.

56. *See* U.S. ENERGY INFORMATION ADMIN., ANN. REP. TO CONGRESS 103 (1978), which reports that more than 30 billion tons of demonstrated reserves having less than 1% sulfur content by weight can be found in states east of the Mississippi. Coal having a heat content of 10,000 BTU per pound (probably a conservative estimate for most eastern bituminous coals) and less than 1% sulfur content by weight would produce considerably less than 0.55 pounds of SO_2, even if scrubbed at only 85% removal efficiency.

57. For 1975 production figures, *see* NATIONAL COAL ASS'N, COAL FACTS 80 (1978–79) (bituminous coal only). Reserve data appear in U.S. ENERGY INFORMATION ADMIN., *supra* note 56, at 103.

58. *See* p. 35, *supra*.

59. Coal industry analysts insisted that the typical utility would buy

coal that would require no more than 85% scrubbing despite EPA's finding that 90% reduction was "achievable." On this argument, conservative utility buying practices would effectively preclude coals which required 85%–90% scrubbing to meet the 0.55 standard. It was only by making this assumption that the NCA could report that an 0.55 ceiling would preclude from 75% to 100% of the coal produced in selected market areas. If one assumed that utilities would buy the cheapest coal that would meet 0.55 by 90% scrubbing, then the NCA itself admitted that preclusion would range from 17% to 45%, rather than 75% to 100%.

The NCA produced some surveys purporting to demonstrate widespread conservatism in utility buying practices. These surveys, however, are inconsistent with EPA experience, *see* Memorandum from Walter C. Barber to Barbara Blum (Dec. 6, 1979) *reprinted in* 10 ENVIR. REP. (BNA) 1872, 1873 (1980) (most utilities have interpreted averaging provisions in SIPs in the least conservative manner possible), as well as the more general arguments developed at pp. 70–72, *supra*. We will not explore this matter further, however, since the text's discussion reveals even more serious flaws in the NCA approach to the preclusion issue.

60. *See* NRDC & EDF Comments, *supra* note 7, at IV–6 (advocacy of 95% scrubbing with ceilings from 0.4 to 0.12 lb./MBTU.)

61. *See, e.g.*, 9 ENVIR. REP. (BNA) 2100 (1979) (statement of George Freeman).

62. *See* Minutes of April 5, 1979 Meeting Concerning Coal Reserve Data and Application of Physical Coal Cleaning Credits, Attachment 8 (EPA Docket No. OAPQS–78–I, Item No. IV–E–11). The NCA survey included reserves from Ohio, Illinois, Indiana, northern West Virginia, and western Kentucky, and was based on an assumption that scrubbers would remove only 85% of the sulfur of any coal purchased by a utility; for the significance of this assumption *see* note 59, *supra*.

63. The NCA survey ignored southern West Virginia and eastern Kentucky. In these two regions, 92% of the reserves could meet a low ceiling, even using conservative assumptions. M. WELLS, R. CHAPMAN, R. YEAGER, V. NEREO, & V. MATUCHA, COAL RESOURCES AND SULFUR EMISSIONS REGULATIONS 6–3 (1980) (table 6–1) (Teknekron Report to EPA, Contract No. 68–02–3136) (data showing 92% of coal reserves (weighted average) in these two regions has sulfur content of 3.0 lb. per MBTU or less (0.55 ceiling can be met by using coal containing 3.67 lb. per MBTU or less)). While much of this coal is of sufficiently high quality to command premium prices for use in metallurgical processes, sub-

stantial amounts of affordable, low-sulfur coal exist in these districts of the Appalachian region.

Although the EPA did not emphasize the omission of southern West Virginia and eastern Kentucky, it did counter the NCA study by redefining relevant coal reserves and mining regions. This exercise suggested that a low ceiling precluded no more than 35% of reserves in the Appalachian region (Ohio, northern West Virginia, and Pennsylvania), and no more than 75% in the Midwest (Illinois, Indiana, and western Kentucky). Minutes of April 5, 1979 Meeting, *supra* note 62, Attachment 3 (memorandum from John D. Crenshaw to John Haines (Mar. 30, 1979)). The EPA also challenged the NCA's claim that utilities would only buy coal that could meet the 0.55 limit when scrubbed at 85% efficiency. *See* note 59, *supra*. If utilities responded to the new NSPS by buying the cheapest coal that could be scrubbed at 90% efficiency, precluded reserves would be dramatically reduced in both Appalachian and midwestern regions. *Id*.

64. *See* Memorandum from John D. Crenshaw to Walter C. Barber (May 11, 1979) (EPA Docket No. OAQPS–78–I, Item No. IV–B–72) (using regions surveyed by NCA, plus Pennsylvania).

65. *See, e.g.*, Memorandum from John D. Crenshaw to Walter C. Barber, *supra* note 64.

66. *See* Memorandum from David G. Hawkins to NSPS Docket (Apr. 26, 1979) (EPA Docket No. OAQPS–78–I, Item No. IV–E–13) (Apr. 23 meeting); Memorandum from David G. Hawkins to Walter C. Barber (May 3, 1979) (EPA Docket No. OAPQS–78–I, Item No. IV–E–24) (May 2, 1979 meeting).

Other coal-state congressmen supported Senator Byrd during this period. The EPA docket contains over a dozen communications (by letter) between concerned representatives and the EPA. *See* EPA Docket No. OAQPS–78–I, Item nos. IV–D–750 to 880 (interspersed comments of individual Congressmen).

67. *See* Memorandum from David G. Hawkins to NSPS Docket, *supra* note 66.

68. The retreat to 1.2 is rationalized in a memorandum from John D. Crenshaw to Walter C. Barber, *supra* note 64. The technical analysis underlying the Crenshaw memorandum was developed by the May 2 White House meeting, where it was distributed to attendees. *See* Memorandum from David G. Hawkins to Walter C. Barber, *supra* note 66 (enclosing distributed material). The docket does not indicate whether the administrator expressed his decision to retreat to a 1.2 standard at that time. EPA has since reported that the administrator had already

concluded that "anything more than minimal preclusion" would be inconsistent with the purposes of the 1977 amendments. 45 Fed. Reg. 8215 (1980).

69. The 1971 NSPS did not explicitly state the period of time during which an old plant was required to maintain a 1.2 emission ceiling. However, it did require a new plant to pass a three-hour performance test. *See* 40 C.F.R. § 60.45(g)(2)(i) (1979). If a three-hour test had been adopted in 1979, the regulation would have been far more restrictive than the one actually established—which only requires plants to meet a 1.2 monthly average. 40 C.F.R. § 60.43a(c) (1979). The 50% figure presented in the text is based on interviews with experts both in and out of the agency.

70. Memorandum from John D. Crenshaw to Walter C. Barber, *supra* note 64, at 6.

71. For a clear statement of many of the objections raised by DOE to full scrubbing written by an active participant in the rule making, *see* Badger, *New Source Standard for Power Plants I: Consider the Costs*, 3 HARV. ENVT'L L. REV. 48 (1979).

72. *See* An Alternative Standard Keyed to Wet and Dry Scrubber Technology (unsigned memorandum) (Apr. 16, 1979) (on file with authors). We do not mean to suggest that the dry scrubber had been entirely ignored at earlier points in the proceedings. Thus, the EPA's initial September rule-making notice made a brief mention of the new technology and some people advocated the use of dry scrubbing at EPA hearings held in December 1978. *See* 45 Fed. Reg. 8,216 (1980). Moreover, middle-level bureaucratic interest in dry scrubbing options is observable as early as mid-February 1979. Memorandum of Feb. 16, 1979 (on file with authors). It was only in April, however, that a high-level decision was made to conduct modeling runs based on newly-constructed cost estimates for the dry scrubber. It is this last-minute reliance on the sketchiest of cost estimates that reveals the desperation of the EPA's effort to reconcile administrative expertise to the needs of special interest politics.

73. In basic concept, dry scrubbing is very similar to wet scrubbing. *See* pp. 15–16, *supra*. The chemical reaction in a dry scrubber, however, produces a dry powder rather than a liquid. *See* Dry SO_2 Control 1–2 (unsigned, undated memorandum) (supplied to authors in May 1978 by members of EPA Office of Research and Development) (on file with authors), *reprinted in part in* Memorandum from Walter C. Barber to the file (May 24, 1979) (EPA Docket No. OAQPS–78–I, Item. No. IV–E–21). In practice, dry scrubbers promise to be mechanically

simpler, to consume less water (an important consideration in the West), and to create fewer sludge handling problems than wet scrubbers. *See id.* at 14; An Alternative Standard Keyed to Wet and Dry Scrubber Technology, *supra* note 72, at 1.

74. *See* Dry SO_2 Control, *supra* note 73, at 5–9.

75. *See, e.g.*, S. REP. No. 1196, 91st Cong., 2d Sess. 417 (1970) (advocating frequent revision of NSPS to provide incentives for constant improvement of control technology).

76. The emission ceilings given in the text assume an EPA practice of measuring compliance on the basis of monthly average discharge. They translate into equivalent annual averages of roughly 0.5 pounds and 1.0 pound of SO_2/MBTU, respectively.

77. The modeling results actually shown White House aides seem to have differed slightly from those published later in more formal documents. *See* Memorandum from Walter C. Barber to the file, *supra* note 73 (table 3) (discussing May 1, 1979 meeting with White House aides).

78. *See* 44 Fed. Reg. 33,582 (1979). The nation's first three full-scale dry scrubbers were still being installed in 1979. *Id.*

79. Compare 37 Fed. Reg. 5769 (1972) (estimated scrubber costs of \$30/KW or \$45/KW in 1978 dollars) with ICF III, *supra* note 4, at 13–21 (estimated scrubber costs as high as \$90–\$113/KW). Recall that the critical contribution of the dry scrubber in resolving EPA's dilemma was not the technology's ability to remove SO_2 but its ability to do so at low cost.

CHAPTER 7

1. Sierra Club v. Costle, No. 79-1565 (D.C. Cir., filed June 11, 1979). EPA denied requests for reconsideration of the standards in February 1980. *See* 45 Fed. Reg. 8210 (1980).

2. *See* R. Stewart, *The Reformation of American Administrative Law*, 88 HARV. L. REV. 1669, 1678–80 (1975).

3. *Cf.* Blumrosen, *Toward Effective Administration of New Regulatory Statutes*, 29 ADMIN. L. REV. 87, 89–100 (1977) (agency administrators should be required to construe reform legislation broadly).

4. Many of the Clean Air Act's sections carrying statutory deadlines fall into this category. *See, e.g.*, 42 U.S.C. § 7409(d)(1) (Supp. II 1978) (duty to review NAAQS's at least once every five years); *id.* § 7411(b)(1)(B) (duty to review NSPSs); *id.* § 7422 (duty to review health effects of certain unregulated pollutants); *id.* § 7408(c) (duty to review air quality criteria and revise criteria for NO_2).

5. *See, e.g.*, 15 U.S.C. § 78g(a)–(b) (1976) (Federal Reserve Board may

set margin requirements for securities purchases lower or higher than congressional standard if appropriate in light of general credit situation).

6. *See, e.g.*, 29 U.S.C. § 655(a) (1976) (requiring Secretary of Labor to promulgate "national consensus standard" or "established Federal standard" as occupational safety or health standard "unless he determines that the promulgation of such a standard would not result in improved safety or health for specifically designated employees"); 15 U.S.C. § 2605(e) (1976) (administrator shall ban polychlorinated biphenyl if not totally enclosed, unless he finds that use will not present unreasonable risk of injury to health or environment).

7. *Cf.* 42 U.S.C. § 7409(c) (Supp. II 1978) (requiring promulgation of three-hour standard for nitrogen oxide unless the administrator finds "no significant evidence that such a standard for such a period is requisite to protect public health").

8. *See, e.g.*, H.R. REP. No. 1013, 93d Cong., 2d Sess., *reprinted in* [1974] U.S. CODE CONG. & AD. NEWS 3281, 3292 (voicing House approval of Judge Leventhal's intervention in International Harvester v. Ruckelshaus, 478 F.2d 615 (D.C. Cir. 1973). As explained in the next paragraph of the text, we understand the principle of full inquiry to be a generalization of the approach taken by Leventhal in the *Harvester* case.

9. 478 F.2d 615 (D.C. Cir. 1973). *See* South Terminal v. EPA, 504 F.2d 646, 665–67 (1st Cir. 1974) (setting aside EPA-promulgated transportation control plan for Boston area due to insufficient justification of standards).

10. *See* Clean Air Act § 202(b)(5)(D)(i)–(iv), formerly codified at 42 U.S.C. § 1857f–1(b)(5)(D)(i)–(iv) (1970) (amended 1974).

11. *See* 478 F.2d 624, 633–41 (D.C. Cir. 1973).

12. *See* pp. 29–30, 50–54, *supra*.

13. In the words of Justice Frankfurter, "[l]oose judicial reading makes for loose legislative writing." Frankfurter, *Some Reflections on the Reading of Statutes*, 47 COLUM. L. REV. 527, 545 (1947).

Although we do not wish to approve the mechanistic approach to textual analysis recently favored by the District of Columbia Circuit in Alabama Power Co. v. Costle, 10 ENVT'L L. REP. 20,001, 20,036–37 (D.C. Cir. 1979), the decision does represent a turn away from casual reliance on low-level legislative history; to that extent it supports the principle presented here. In *Alabama Power*, the court supported the EPA's decision to measure emissions on a plantwide basis rather than treating each smokestack as if it were a separate polluter. By placing an entire plant under a single "bubble," the agency relieved the polluter of the need to clean up each smokestack equally if other measures would reach the plantwide target more cheaply. While we can only applaud the court's conclusion, *see* note 6, chap. 8, *infra*, its reasoning left a good deal to be desired. Rather than relate its decision to the general policy of encouraging cost-effective control measures, the court solved its problem by an exceedingly literal treatment of the particular words of the particular subsection under dispute.

14. *See, e.g.*, 42 U.S.C. § 7401(b)(1) (Supp. II 1978) (purpose of the act is to "protect and enhance the quality of the Nation's air resources").

15. This is not the place for an exploration of the constitutional limits placed on congressional efforts to seal off certain regions of the country from nationwide competition. Although the great case of Cooley v. Board of Port Wardens, 53 U.S. 299 (1851) is best read to suggest that Congress cannot legitimate gross forms of regional protectionism, later decisions have given Congress great latitude in this regard. *See* L. TRIBE, AMERICAN CONSTITUTIONAL LAW 402–03 (1978). Cases like the one discussed here, however, suggest the need for a thoughtful reassessment of constitutional principles governing congressional power under the commerce clause.

16. *See generally*, Bruff, *Presidential Power and Administrative Rulemaking*, 88 YALE L.J. 451, 498–508 (1979).

17. *See* 42 U.S.C. § 7417(c) (Supp. II 1978).

18. *Cf.* Scenic Hudson Preservation Conf. v. Federal Power Comm'n (I), 354 F.2d 608 (2d Cir. 1965), *cert. denied*, 384 U.S. 941 (1966).

19. *Cf.* Portland Cement v. Ruckelshaus, 486 F.2d 375, 387 (D.C. Cir. 1973) (EPA should consider cost-benefit studies submitted to it); Kennecott Copper Corp. v. EPA, 462 F.2d 846, 850 (D.C. Cir. 1972) (requirement that EPA supply sufficiently detailed explanation of basis of promulgated standard for court to review). The agency, in its promulgation, offered no response even to the crude exposure index developed by RARG. *See* 44 Fed. Reg. 33, 602–08 (1979), and pp. 93–94, *supra*.

20. 43 Fed. Reg. 57,857 (1978).

21. Participants in the rule making have argued that Senator Byrd's meeting with agency officials was improper both because it constituted a prohibited ex parte contact, *see* Home Box Office, Inc. v. FCC, 567 F.2d 9, 57 (D.C. Cir. 1977), and because it allowed the Senator to influence the rule making by use of political pressure. *See* D.C. Federation of Civic Ass'ns v. Volpe, 459 F.2d 1231, 1245–46 (D.C. Cir. 1972)

(invalidating administrator's decision based partially on political pressure). For EPA's response to these arguments, see 45 Fed. Reg. 8214–15 (1980).

22. In fact, congressmen questioned EPA officials on several occasions, sometimes quite sharply, about NSPS during hearings that were held while the rule making was still in progress. *See, e.g., Executive Branch Review of Environmental Regulations: Hearings before the Subcomm. on Environmental Pollution of the Senate Comm. on Environment and Public Works*, 96th Cong., 1st Sess. 252–53 (1979) (statement of Douglas Costle) (answering a question critical of protecting eastern coal); *Cost of Government Regulations to the Consumer: Hearings before the Subcomm. for the Consumer of the Senate Comm. on Commerce, Science, and Transportation*, 95th Cong., 2d Sess. 323–26 (1978) (statement of Walter C. Barber) (answering questions critical of sludge produced by scrubbers). And there is nothing the courts could do to prevent Congress from closing these sessions to the public if the committees believed this to be in the public interest.

Courts have traditionally been reluctant to reverse administrative decisions on the basis of exchanges between administrative officials and congressmen. *See, e.g.*, American Pub. Gas Ass'n v. Federal Power Comm'n, 567 F.2d 1016, 1067–69 (D.C. Cir. 1977); Gulf Oil Corp. v. Federal Power Comm'n, 563 F.2d 588, 610–11 (3d Cir. 1977).

23. EPA contended both before and after the meeting with Senator Byrd that its actions were designed to be consistent with *Home Box Office*, 567 F.2d at 57. *See Executive Branch Review of Environmental Regulations: Hearings before the Subcomm. on Environmental Pollution of the Senate Comm. on Environment and Public Works, supra* note 22 at 242 (statement of Douglas Costle); 45 Fed. Reg. 8214–15 (1980) (denying petitions for reconsideration). The fact that EPA recorded a significant number of ex parte contacts in the rule-making docket tends to support this assertion.

24. It is also rumored that Senator Byrd threatened to vote against the SALT II treaty if EPA did not retreat to a 1.2 ceiling. We have found no hard evidence to support this rumor and our analysis makes it unnecessary for the court of appeals to give it any weight. Even if Byrd did not make this threat and restricted himself to arguments on behalf of eastern coal, the agency was wrong to give way before him. Until eastern coal succeeds in revising the statutory text, its interests are not entitled to special prominence in the administrator's comprehensive analysis of the costs of the NSPS decision.

25. *Cf.* Kennecott Copper Corp. v. EPA, 462 F.2d 846, 850 (D.C. Cir. 1972) (requiring sufficiently reasoned explanation of basis of standards for informed judicial review).

CHAPTER 8

1. As we noted earlier, eastern coal reserves include significant amounts of low-sulfur coal. *See* pp. 97–99, *supra*. Because these reserves make up only a small portion of all eastern coal, however, we have used the term "eastern coal" in this section as a shorthand reference to eastern high-sulfur coal producers.

In contrast to their high-sulfur brethren, eastern low-sulfur producers had the greatest possible interest in battling *against* forced scrubbing—since this would place their product at a great advantage over both western coal (which could be railroaded east only at great expense) as well as eastern high-sulfur coal (which required expensive scrubbing to meet the 1.2 NSPS). The failure of eastern low-sulfur interests to mount a significant political effort remains a puzzling fact that we have been unable to explain to our own satisfaction.

2. Our concern with interest group politics is presently unfashionable among professional political scientists—who, after a generation in which they could think of little else, see D. TRUMAN, THE PROCESS OF GOVERNMENT (2d ed. 1971), have more recently consigned the issue to the periphery of concern. Thus, the most salient work on interest groups remains the study by R. BAUER, I. POOL, & L. DEXTER, AMERICAN BUSINESS AND PUBLIC POLICY (1963), which found interest group activity surprisingly insignificant in connection with tariff legislation of the 1950s. Whatever the facts about the tariff, it is wrong to generalize a single case study into a universal truth. When billions of dollars can be made or lost by manipulating low-visibility legislation, it is naive to assume that businessmen will never organize to take advantage of such large opportunities. For a suggestive reappraisal, emphasizing the subtle ways in which interest groups may exploit congressional ignorance, *see* R. Smith, Lobbying Influence in Congress (unpublished paper presented at the 1979 meetings of the American Political Science Association). Rather than ignore the dangers of interest group manipulation of Congress, the task is to design institutions that will ameliorate, if not eliminate, the abuses of faction. *See* J. Madison, *Federalist Ten*, in THE FEDERALIST PAPERS 77–84 (C. Rossiter ed. 1961).

3. *See* Marcus, *Environmental Protection Agency*, in THE POLITICS OF REGULATION 267–74 (J. Wilson ed. 1980); Ingram, *The Political Rational-*

ity of Innovation: The Clean Air Act Amendments of 1970 in APPROACHES TO CONTROLLING AIR POLLUTION 12–64 (A. Friedlaender ed. 1978).

4. For reviews of the present learning, *see* Sabatier, *Social Movements and Regulatory Agencies: Toward a More Adequate—and Less Pessimistic —Theory of "Clientele Capture,"* 6 POLICY SCIENCES 301 (1975); POSNER, *Theories of Economic Regulation*, 5 BELL J. OF ECON. & MGMENT SCI. 335 (1974).

5. *See, e.g., Clean Air Act Amendments of 1977: Hearings before the Subcomm. on Environmental Pollution of the Senate Comm. on Environment and Public Works*, 95th Cong., 1st Sess. vi–xxi (1977) (summary of testimony prepared by Congressional Research Service); *Clean Air Act Amendments of 1977: Hearings before the Subcomm. on Health and the Environment of the House Comm. on Interstate and Foreign Commerce*, 95th Cong., 1st Sess., 240–62 (statement of Carl M. Shy, American Lung Association) (focused on health effects of various pollutants); *id.* at 1250–1329 (statements of representatives of numerous industry organizations largely focusing on question of act's proper ends).

6. Section 173 of the act permits continued economic growth in areas that have not attained clean air targets by allowing the construction of a new plant on condition that the plant obtain offsetting emission reductions from existing plants in the area. 42 U.S.C. § 7503(1)(A) (Supp. II 1978). This new provision represents an important step in the right direction—evidencing congressional recognition that the important thing is to clean the air and that it does not matter whether old or new plants are doing the cleaning. Unfortunately, however, Section 173 does not carry this ecologically sound position to its logical conclusion. While requiring a new plant to offset its emissions by inducing others to cut back on their discharges, the statute *also* requires the new plant to limit its own discharge to the "lowest achievable emission rate" (LAER). Thus, if a new coal burner's LAER is 1,000 tons of SO_2, it cannot discharge 3,000 tons *even if it is cheaper to induce an old coal burner to reduce its load by 2,000 tons* than install LAER technology.

By adding a means-oriented LAER requirement to the ends-oriented offset system, Section 173 forfeits many of the efficiency gains achievable under the offset. The section should be viewed as a transitional stage in the evolution of the act from a means-oriented to an ends-oriented focus, and steps should be taken at the next congressional revision in the early 1980s to liberate offset from the losses imposed by the LAER requirement. For a proposal, *see* note 38, chap. 5.

7. Most recently, the Court of Appeals for the District of Columbia has sustained the EPA's decision to allow offsets between different facil-

ities within the same plant as part of the agency's "bubble" concept. *See* Alabama Power Co. v. Costle, 10 Envt'l. L. Rep. 20,001, 20,036–37 (D.C. Cir. 1979).

Prior to *Alabama Power Co.*, the majority of a panel of the same circuit concluded that EPA had the authority to authorize offsets between facilities of the same plant, even though it had not used that authority properly in the case before it. ASARCO, Inc. v. EPA, 578 F.2d 319, 330–31 (D.C. Cir. 1978) (Leventhal, J. concurring and MacKinnon, J., dissenting).

Judge Leventhal's opinion in *ASARCO* is an instructive example of the principles outlined in the text. Judge Leventhal's approach first required the administrator to use the agency's expertise to identify situations where innovative plans like an offset policy would further underlying statutory goals, such as reducing emissions from modified sources in a way that best takes cost into account. In *ASARCO* itself, Judge Leventhal found that EPA had not made a full enough inquiry into the merits of an offset policy to justify its use. *Id.* at 331. More important, however, Judge Leventhal's opinion found textual authority, under Section 111(b)(2), for the use of offsets between facilities of the same plant where sufficient study had occurred and thus offers support for more far-reaching innovations.

8. Although their proposals are different, the need for more operational targets in the Clean Air Act is also emphasized by J. Krier & E. Ursin, Pollution and Policy 327–45 (1977).

9. For a good history of the evolution of agency awareness of sulfates during the past decade, *see* R. Tobin, The Social Gamble: Determining Acceptable Levels of Air Quality 127–576 (1979); *see also* pp. 60–64, *supra*.

10. *See* pp. 60–64, *supra*.

11. *See*, for example, the unhappy history of the EPA's CHESS studies, recounted by Tobin, *supra* note 9, at 102–06, 139–43.

12. For a recent discussion of the increasingly popular concept of the regulatory budget, *see Regulatory Budgeting and the Need for Cost-Effectiveness in the Regulatory Process: Hearing before the Joint Economic Comm.*, 96th Cong., 1st Sess. (1979).

13. Congress in 1977 established a National Commission on Air Quality to advise it before the act returns for revision. Clean Air Act Amendments of 1977, § 313, 42 U.S.C. § 7623 (Supp. II 1978). As a result of the leisurely appointment of commissioners, the panel is off to a slow start, and it is too soon to assess its work. For a general discussion of the dangers of leaving the long-range planning function to *ad hoc*

bodies like the commission *see*, B. ACKERMAN, et al., THE UNCERTAIN SEARCH FOR ENVIRONMENTAL QUALITY, *supra* note 34, chap. 5, at 73–77.

14. For an intelligent discussion of the need to design federal institutions that can sensitively operate on the regional level, *see* Roberts, *Organizing Water Pollution Control: The Scope and Structure of River Basin Authorities*, 19 PUBLIC POLICY 75 (1971); Roberts, *River Basin Authorities: A National Solution to Water Pollution*, 83 HARV. L. REV. 1554 (1970).

15. *See, e.g., Regulatory Budgeting and the Need for Cost-Effectiveness in the Regulatory Process: Hearing before the Joint Economic Comm.*, *supra* note 12; S. COMM. ON GOVERNMENTAL AFFAIRS, STUDY ON FEDERAL REGULATION, 95th Cong., 1st Sess. (1977) 168–69 (Comm. Print 1979).

16. *See, e.g.*, REGULATORY REFORM: MESSAGE FROM THE PRESIDENT, H.R. DOC. No. 80, 96th Cong., 1st Sess. (1979).

17. *See, e.g.*, Breyer, *Analyzing Regulatory Failure: Mismatches, Less Restrictive Alternatives, and Reform*, 92 HARV. L. REV. 549 (1979); C. DeMuth, Regulatory Costs and the "Regulatory Budget," (Dec. 1979) (unpublished faculty project on regulation, John F. Kennedy School of Government, Harvard University). For a small sampling of the many projects now in progress, *see* E. BARDACH & R. KAGAN, THE INSPECTORATE (forthcoming, provisional title); R. Litan & W. Nordhaus (forthcoming work on regulatory reform).

INDEX

Acid rain, 66

Administrative law. *See* Judicial review; New Deal agency; Post New Deal agency

Agency forcing statutes, 3, 26–27, 114, 124–28; ends-forcing, 9, 11, 21, 57, 109, 122–28; means forcing, 11, 21, 122, 123, 124; and principle of full inquiry, 104–07; and agenda-forcing, 105, 109, 114; and solution forcing, 105, 109, 114; and legislative history, 108–09; and executive coordination, 110–11. *See also* Legislation; New Deal agency; Post New Deal agency

Agricultural production: and acid rain, 66

Air office. *See* Office of Air, Noise, and Radiation (EPA)

Air pollution: cost of abatement, 129, 160. *See also* Carbon dioxide; Greenhouse effect; Particulate matter; Sulfates; Sulfur dioxide

Air quality: in eastern U.S., 61–62, 63, 65–67, 151, 157–58; in western U.S., 75, 77, 89, 95–96, 161, 163, 168, 169, 171

Air quality criteria: for sulfur oxides, 61–62

Air quality monitoring. *See* Monitoring

Andrus, Cecil, 112

Aquatic life: as affected by acid rain, 66

Automobile emissions, 27; 1977 deadline for, 42; moves to delay compliance, 48; *International Harvester*, 107

Ayres, Richard, 36–37, 142, 146–47

BACT. *See* Best available control technology

BTU. *See* British thermal unit

Bayh, Birch: support of § 125, 45

Best available control technology (BACT), 147; EPA endorsement of, 148

Brecher, Joseph, 36

British thermal unit (BTU): definition of, 138

Business roundtable, 92

Byrd, Robert, 101, 112; opposition to low-ceiling option, 100; and SALT II, 179

CEA. *See* Council of Economic Advisors

CO_2. *See* Carbon dioxide

COWPS. *See* Council on Wage and Price Stability

Carbon dioxide: greenhouse effect produced by, 129

44, 143; effect of NSPS on consumption, 166

O'Leary, John: opposition to full scrubbing, 87

PSD. *See* Prevention of significant deterioration

Pareto optimality, 88–89

Partial scrubbing, 87–90; advocated by RARG, 93; legal support for, 106–15

Particulate matter: produced by power plants, 129

Percentage reduction requirement: legislative creation of, 51–54; compared to marketlike systems, 69; compared to lower ceiling, 69–70; enforcement problems of, 70–72; and visibility, 77–78; and strip-mining, 78. *See also* Full scrubbing

Physical coal cleaning: description of, 15, 136; and 1971 NSPS, 16–17; relative cost of, 67, 156–57; emission reductions attainable by, 67–69, 161

Planning office. *See* Office of Planning and Management

Political science: and institutional reform, 180

Post New Deal agency, 1, 7–12, 132; and congressional failures, 54–58; recommended changes, 57–58; failures of, 68, 103; judicial review of, 104–15. *See also* Agency-forcing statutes

Power plants: as source of pollutants, 2, 23, 66–68; reliance on coal, 2, 35, 97–99; new source performance standards for, 2, 79–103 passim; limitations on existing plants, 10;

compliance with Clean Air Act, 28, 140; percentage reduction requirement for, 51–54; captive mines, 74

Preclusion: of high sulfur coal, 19, 32–33, 37, 97–101, 172–75; environmentalists' concern for, 36, 98; and administrative procedure, 113

Prevention of significant deterioration (PSD), 134; history of, 28–29; influence on 1977 amendments, 28–33, 36, 39–41, 48–49, 56; 1975-76 technical studies of, 33–35

Public-interest lobbyists. *See* Environmental groups, Natural Resources Defense Council; Sierra Club

Public utilities. *See* Power plants; Utility industry

RARG. *See* Regulatory Analysis Review Group

Randolph, Jennings: support for 1977 amendments, 50; support of § 125, 149

Rayleigh scattering, 75

Regional protectionism, 11, 31; § 125, 44–47; constitutional limits, 178. *See also* Clean Air Act: —1977 amendments, § 125; Federalism

Regulatory Analysis Review Group (RARG): membership and duties of, 91; value of, 92–93; recommendations of, 93–94; and judicial review, 111

Regulatory budget, 126, 182–83

Research. *See* Scientific research

Research and Development, EPA Office of, 101